NATEF Standards Job Sheets

Engine Performance (A8)

Second Edition

Jack Erjavec

THOMSON
DELMAR LEARNING

Australia Canada Mexico Singapore Spain United Kingdom United States

THOMSON
DELMAR LEARNING

NATEF Standards Job Sheets

Engine Performance (A8)
Second Edition

Jack Erjavec

Vice President, Technology and Trades SBU:
Alar Elken

Editorial Editor:
Sandy Clark

Senior Acquisitions Editor:
David Boelio

Development Editor:
Matthew Thouin

Marketing Director:
David Garza

Channel Manager:
William Lawrensen

Marketing Coordinator:
Mark Pierro

Production Director:
Mary Ellen Black

Production Editor:
Toni Hansen

Art/Design Specialist:
Cheri Plasse

Technology Project Manager:
Kevin Smith

Editorial Assistant:
Andrea Domkowski

Printed in United States of America
1 2 3 4 5 XXXX 09 08 07 06 05

For more information contact Thomson Delmar Learning, Executive Woods, 5 Maxwell Drive, Clifton Park, NY 12065-8007.
Or find us on the World Wide Web at http://www.delmarlearning.com.

ISBN: 1-4180-2081-8

NOTICE TO THE READER

CONTENTS

Preface v

Engine Performance Systems 1

Basic Engine Performance Theory 1
Safety 16
Hazardous Materials and Wastes 20
NATEF Task List for Engine Performance 22
Definition of Terms Used in the Task List 25
Engine Performance Tools and Equipment 27

Cross-Reference Guide 45

Job Sheets 47

Job Sheet 1 Filling Out a Work Order 49
Job Sheet 2 Verifying the Condition of and Inspecting an Engine 53
Job Sheet 3 Gathering Vehicle Information 63
Job Sheet 4 Interpreting Identification Numbers 67
Job Sheet 5 Using a Vacuum Gauge 69
Job Sheet 6 Conduct a Cylinder Power Balance Test 71
Job Sheet 7 Perform a Cylinder Cranking Compression Test 73
Job Sheet 8 Perform an Engine Running Compression Test 75
Job Sheet 9 Perform a Cylinder Leakage Test 79
Job Sheet 10 Scope Testing an Ignition System 83
Job Sheet 11 Using an Exhaust Gas Analyzer 87
Job Sheet 12 Engine Temperature Check 89
Job Sheet 13 Test Cooling System 93
Job Sheet 14 Valve Timing Check 95
Job Sheet 15 Retrieve Codes from the Computer of an Engine Control System 97
Job Sheet 16 Conduct a Diagnostic Check on an Engine Equipped with OBD II 99
Job Sheet 17 Interpreting Codes from the Computer of an Engine Control System 103
Job Sheet 18 No-Code Diagnostics 107
Job Sheet 19 Using a Scan Tool on Engine Control Systems 111
Job Sheet 20 Checking Common Sensors 115
Job Sheet 21 Electronic Service Information 123
Job Sheet 22 Diagnosis of Power Supply and Ground Circuits 125
Job Sheet 23 Handling Static Sensitive Devices 129
Job Sheet 24 Diagnosing Related Systems 131
Job Sheet 25 Using a Scan Tool to Test Actuators 137
Job Sheet 26 No-Start Diagnosis 139
Job Sheet 27 Visually Inspecting an Electronic Ignition System 143
Job Sheet 28 Individual Component Testing 147

Job Sheet 29	Testing an Ignition Coil	153
Job Sheet 30	Setting Ignition Timing	155
Job Sheet 31	Diagnosing EFI Systems	157
Job Sheet 32	Inspecting the Fuel System	165
Job Sheet 33	Checking Fuel for Contaminants	167
Job Sheet 34	Checking Fuel Pumps	169
Job Sheet 35	Checking EFI Fuel Delivery Circuit	173
Job Sheet 36	Removing a Fuel Filter on an EFI Vehicle	175
Job Sheet 37	Performing a Scan Tester Diagnosis of an Idle Air Control Motor	177
Job Sheet 38	Servicing a Throttle Body	179
Job Sheet 39	Inspect, Clean, and Test Fuel Injectors	181
Job Sheet 40	Visually Inspect an EFI System	187
Job Sheet 41	Checking Idle Speed	189
Job Sheet 42	Checking Fuel Delivery Circuits	193
Job Sheet 43	Inspect Exhaust System	195
Job Sheet 44	Test a Catalytic Converter for Efficiency	199
Job Sheet 45	Checking Boost Systems	203
Job Sheet 46	Check the Operation of a PCV System	207
Job Sheet 47	Check the Operation of an EGR Valve	211
Job Sheet 48	Air Injection System Diagnosis and Service	215
Job Sheet 49	Evaporative Emission Control System Diagnosis	221
Job Sheet 50	Adjusting Valves	225
Job Sheet 51	Replace a Timing Belt on an OHC Engine	229
Job Sheet 52	Servicing a Thermostat	233
Job Sheet 53	Checking Cooling Fans	237
Job Sheet 54	How to Repair and Replace Damaged Threads	241
Job Sheet 55	Perform Oil and Filter Change	245
Job Sheet 56	Using an Oxy-Acetylene Torch	249
Job Sheet 57	Servicing the Internal Combustion Engine on a Hybrid Vehicle	255

PREFACE

The automotive service industry continues to change with the technological changes made by automobile and tool and equipment manufacturers. Today's automotive technician must have a thorough knowledge of automotive systems and components, good computer skills, exceptional communication skills, good reasoning, the ability to read and follow instructions, and above average mechanical aptitude and manual dexterity.

This new edition, like the last, was designed to give students a chance to develop the same skills and gain the same knowledge that today's successful technician has. This edition also reflects the changes in the guidelines established by the National Automotive Technicians Education Foundation (NATEF), as of July 2005.

The purpose of NATEF is to evaluate technician training programs against standards developed by the automotive industry and recommend qualifying programs for certification (accreditation) by ASE (National Institute for Automotive Service Excellence). Programs can earn ASE certification upon the recommendation of NATEF. NATEF's national standards reflect the skills that students must master. ASE certification through NATEF evaluation ensures that certified training programs meet or exceed industry-recognized, uniform standards of excellence.

At the expense of much time and many minds, NATEF has assembled a list of basic tasks for each of their certification areas. These tasks identify the basic skills and knowledge levels that competent technicians have. The tasks also identify what is required for a student to start a successful career as a technician.

Most of the content in this book are job sheets. These job sheets relate to the tasks specified by NATEF. The main considerations during the creation of these job sheets were student learning and program certification by NATEF. Students are guided through standard industry accepted procedures. While they are progressing, they are asked to report their findings as well as offer their thoughts on the steps they have just completed. The questions asked of the students are thought provoking and require students to apply what they know to what they observe.

The job sheets were also designed to be generic. That is, whenever possible, the tasks can be performed on any vehicle from any manufacturer. Also, completion of the sheets does not require the use of specific brands of tools and equipment; rather students use what is available. In addition, the job sheets can be used as a supplement to any good textbook.

Also included are description and basic use of the tools and equipment listed in NATEF's standards. The standards recognize that not all programs have the same needs, nor do all programs teach all of the NATEF tasks. Therefore, the basic philosophy for the tools and equipment requirement is that the training should be as thorough as possible with the tools and equipment necessary for those tasks.

Theory instruction and hands-on experience of the basic tasks provide initial training for employment in automotive service or further training in any or all of the specialty areas. Competency in the tasks indicates to employers that you are skilled in that area. You need to know the appropriate theory, safety, and support information for each required task. This should include identification and use of the required tools and testing and measurement equipment required for the tasks, the use of current reference and training materials, the proper way to write work orders and warranty reports, and the storage, handling, and use of Hazardous Materials as required by the

'Right to Know Law', and federal, state, and local governments.

Words to the Instructor: I suggest you grade these job sheets based on completion and reasoning. Make sure the students answer all questions. Then look at their reasoning to see if the task was actually completed and to get a feel for their understanding of the topic. It will be easy for students to copy others' measurements and findings, but each student should have their own base of understanding and that will be reflected in their explanations.

Words to the Student: While completing the job sheets, you have a chance to develop the skills you need to be successful. When asked for your thoughts or opinions, think about what you observed. Think about what could have caused those results or conditions. You are not being asked to give accurate explanations for everything you do or observe. You are only asked to think. Thinking leads to understanding. Good technicians are good because they have a basic understanding of what they are doing and of why they are doing it.

Jack Erjavec

ENGINE PERFORMANCE SYSTEMS

To prepare you to learn what you should learn from completing the job sheets, some basics must be covered. This discussion begins with an overview of engine performance systems. Emphasis is placed on what they do and how they work. This includes the major components and designs of engine systems and their role in the efficient operation of engines of all designs.

Preparing to do something on an automobile would not be complete if certain safety issues were not addressed. This discussion covers those things you should and should not do while working on engine performance systems. Included are proper ways to deal with hazardous and toxic materials.

NATEF's task list for Engine Performance certification is also given with definitions of some of the terms used to describe the tasks. This list gives you a good look at what the experts say you need to know before you can be considered competent to work on engine performance systems.

Following the task list are descriptions of the various tools and types of equipment you need to be familiar with. These are the tools you will use to complete the job sheets. They are also the tools NATEF has identified as being necessary for servicing engine performance systems.

Following the tool discussion is a cross-reference guide that shows what NATEF tasks are related to specific job sheets. In most cases there are single job sheets for each task. Some tasks are part of a procedure and when that occurs, one job sheet may cover two or more tasks. The remainder of the book contains the job sheets.

BASIC ENGINE PERFORMANCE THEORY

In order for an engine to run efficiently, complete combustion must take place in each of its cylinders. To have complete combustion, the cylinders must receive the correct amount of fuel mixed with the correct amount of air. Then, that mixture must be shocked with the right amount of heat at the correct time. An engine is fitted with many and various systems that work toward achieving the goal of total combustion.

Ignition Systems

One of the requirements for an efficiently running engine is the correct amount of heat delivered into the cylinders at the right time. This requirement is the responsibility of the ignition system. The ignition system supplies properly timed, high-voltage surges to the spark plugs (Figure 1). These voltage surges cause combustion inside the cylinder.

For each cylinder in an engine, the ignition system has three main jobs. First, it must generate an electrical spark that has enough heat to ignite the air/fuel mixture in the combustion chamber. Second, it must maintain that spark long enough to allow for the combustion of all the air and fuel in the cylinders. Finally, it must deliver the spark to each cylinder so that combustion can

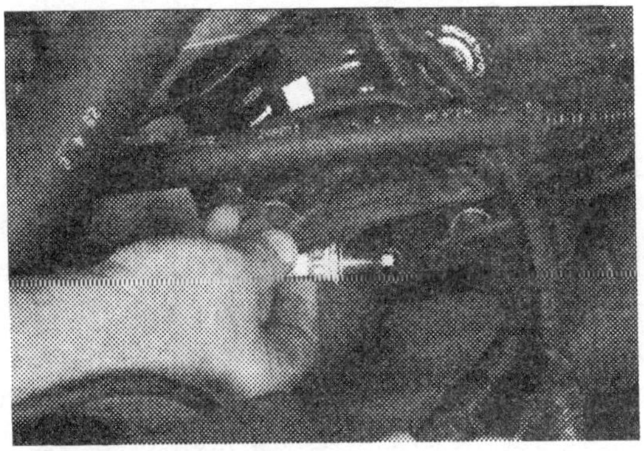

Figure 1 A spark plug.

begin at the right time during the compression stroke of each cylinder.

Ignition Timing

Ignition timing refers to the precise time that spark occurs. If optimum engine performance is to be maintained, the ignition timing of the engine must change as the operating conditions of the engine change.

At higher engine speeds, the crankshaft turns through more degrees in a given period of time. If combustion is to be completed at the correct time, ignition timing must occur sooner or be advanced.

However, air/fuel mixture turbulence increases with rpm. This causes the mixture, inside the cylinder, to burn faster. Increased turbulence requires that ignition must occur slightly later or be slightly retarded. During heavy loads, the mixture burns faster so the ignition timing must again be retarded or combustion will end too soon. Achieving correct ignition timing is the result of careful balancing of engine operating conditions, especially speed and load.

Firing Order

Each cylinder of an engine produces power once every 720 degrees of crankshaft rotation. Each cylinder must have a power stroke at its own appropriate time during the rotation. To make this possible, the pistons and rods are arranged in a precise fashion.

The ignition system must be able to monitor the rotation of the crankshaft and the relative position of each piston in order to determine which piston is on its compression stroke. It must also be able to deliver a high-voltage surge to each cylinder at the proper time during its compression stroke.

Basic Ignition System

All ignition systems consist of two interconnected electrical circuits: a primary (low voltage) circuit and a secondary (high voltage) circuit. Depending on the exact type of ignition system, components in the primary circuit include the battery, ignition switch, ballast resistor or resistance wire (some systems), starting by-pass (some systems), ignition coil primary winding, triggering device, and a switching device or control module. The secondary circuit includes the ignition coil secondary winding, distributor cap and rotor (some systems), ignition (spark plug) cables, and spark plugs.

Primary Circuit Operation

When the ignition switch is in the on position, current from the battery flows through the ignition switch to the primary winding of the ignition coil. From there it passes through some type of switching device and back to ground. The current flow in the ignition coil's primary winding creates a magnetic field. The switching device or control module interrupts this current flow at predetermined times. When it does, the magnetic field in the primary winding collapses. This collapse generates a high-voltage surge in the secondary winding of the ignition coil.

Secondary Circuit Operation

The secondary circuit carries high voltage to the spark plugs. The exact manner in which the secondary circuit delivers these high-voltage surges depends on the system design. Until 1984 all ignition systems used some type of distributor to accomplish this job. However, in an effort to reduce emissions, improve fuel economy, and boost component reliability, most auto manufacturers are now using distributorless or Electronic Ignition (EI) systems.

In a Distributor Ignition (DI) system, high voltage from the secondary winding passes through an ignition cable running from the coil to the distributor. The distributor then distributes the high voltage to the individual spark plugs through a set of ignition cables. The cables are arranged in the distributor cap according to the firing order of the engine. A rotor, driven by the distributor shaft, rotates and completes the electrical path from the secondary winding of the coil to the individual spark plugs.

EI systems have no distributor; instead, spark distribution is controlled by an electronic control unit and/or the vehicle's computer. Instead of a single ignition coil for all cylinders, each cylinder may have its own ignition coil, or two cylinders may share one coil. The coils are wired directly to the spark plug they control. An ignition control module controls the firing order and the spark timing and advance. In many EI systems, a crank sensor located at the front of the crankshaft is used to trigger the ignition system.

Ignition Coil Action

To generate a spark to begin combustion, the ignition system must deliver high voltage to the spark plugs. Because the amount of voltage required to bridge the gap of the spark plug varies with the operating conditions, ignition systems can easily supply 30,000 to 60,000 volts to force a spark across the air gap. Since the battery delivers 12 volts, a method of stepping up the voltage must be used. Multiplying battery voltage is the job of a coil.

The ignition coil is a pulse transformer. It transforms battery voltage into short bursts of high voltage.

A laminated soft iron core is positioned at the center of the ignition coil and an insulated primary winding is wound around this core. The primary winding has approximately 200 turns of heavy wire and the two ends of this winding are connected to the primary terminals on top of the coil. These terminals are usually identified with positive and negative symbols. Enamel-type insulation prevents the primary windings from touching each other. Paper insulation is also placed between the layers of windings.

Secondary coil windings are made of very fine wire. These windings are on the inside of the primary winding. The ends of the secondary winding are usually connected to one of the primary terminals and to the high-tension terminal in the coil tower. The coil assembly is filled with oil through the screw hole in the high-tension terminal. The unit is sealed to protect the windings and to keep the oil inside. The oil helps to cool the coil.

Most coils are not oil-cooled, but are air-cooled instead. These coils are constructed in much the same way except the core is constructed from laminated sheets of iron. These sheets are shaped like the letter "E" and the primary and secondary windings are wound around the center of the E-core.

Spark Plugs

Every type of ignition system uses spark plugs. The spark plugs provide the crucial air gap across which the high-voltage current from the coil flows across in the form of an arc. The three main parts of a spark plug are the steel core, the ceramic core or insulator, which acts as a heat conductor, and a pair of electrodes, one insulated in the core and the other grounded on the shell. The shell holds the ceramic core and electrodes in a gas-tight assembly and has the threads needed for plug installation in the engine. An ignition cable connects the secondary to the top of the plug. Current flows through the center of the plug and arcs from the tip of the center electrode to the ground electrode. The resulting spark ignites the air/fuel mixture in the combustion chamber. Most automotive spark plugs also have a resistor between the top terminal and the center electrode. This resistor reduces radio frequency interference (RFI), which prevents noise on stereo equipment. The resistor, like all other resistances in the secondary, increases the voltage needed to jump the gap of the spark plug.

Spark plugs come in many different sizes and designs to accommodate different engines. To fit properly, spark plugs must be of the proper size and reach. Another design factor that determines the usefulness of a spark plug for a specific application is its heat range. The desired heat range depends on the design of the engine and on the type of driving conditions the vehicle is subject to.

A terminal post on top of the center electrode is the point of contact for the spark plug cable. A ceramic insulator surrounds the center electrode, commonly made of a copper alloy, and a copper and glass seal is located between the electrode and the insulator. These seals prevent combustion gases from leaking out of the cylinder. Ribs on the insulator increase the distance between the terminal and the shell to help prevent electric arcing on the outside of the insulator. The steel spark plug shell is crimped over the insulation and a ground electrode, on the lower end of the shell, is positioned directly below the center electrode. There is an air gap between these two electrodes, and the width of this air gap is specified by the auto manufacturer.

Ignition Cables

Spark plug, or ignition, cables make up the secondary wiring. These cables carry the high voltage from the distributor or the multiple coils to the spark plugs. The cables are not solid wire; instead, they contain fiber cores that act as a resistor in the secondary circuit. They cut down on radio and television interference, increase firing voltages, and reduce spark plug wear by decreasing current. Metal terminals on each end of the spark plug wires contact the spark plug and the distributor cap terminals. Insulated boots on the ends of the cables strengthen the connections as well as prevent dust and water infiltration and voltage loss.

Spark Timing Systems

Electronic switching components can be located inside a separate housing known as an electronic control module or be part of the PCM. Early electronic ignitions still relied on mechanical and vacuum advance mechanisms in the distributor for timing control. Today, timing changes are controlled by the PCM based on inputs from various sensors.

Based on the inputs it receives, the computer makes decisions regarding spark timing and sends signals to the ignition module to fire the spark plugs according to those inputs and according to the programs in its memory.

Crankshaft Position Sensors

The time when the primary circuit must be opened and closed is related to the position of the pistons and the crankshaft. Therefore, the position of the crankshaft is used to control the action of the switching unit.

A number of different types of sensors are used to monitor the position of the crankshaft. These crankshaft position sensors serve as triggering devices and include magnetic pulse generators, metal detection sensors, Hall-effect sensors, and photoelectric sensors.

The mounting location of these sensors depends on the design of the ignition system. All four types of sensors can be mounted in the distributor, which is turned by the camshaft.

Magnetic pulse generators and Hall-effect sensors can also be located on the crankshaft. These sensors are also commonly used on EI ignition systems. Both Hall-effect sensors and magnetic pulse generators can also be used as camshaft reference sensors to identify which cylinder is the next one to fire.

The magnetic pulse or PM generator operates on basic electromagnetic principles. Remember that a voltage is induced when a conductor moves through a magnetic field. The magnetic field is provided by the pick-up unit and the rotating timing disc provides the movement through the magnetic field needed to induce voltage.

The Hall-effect sensor produces a square wave signal that is more compatible with the digital signals required by on-board computers. The operation of a Hall-effect sensor is based on the Hall-effect principle, which states: If a current is allowed to flow through a thin conducting material, and that material is exposed to a magnetic field, voltage is produced.

The heart of the Hall generator is a thin semiconductor layer. Attached to it are two terminals—one positive, and the other negative—that are used to provide the source current for the Hall transformation.

Directly across from this semiconductor element is a permanent magnet. It is positioned so that its lines of flux bisect the Hall layer at right angles to the direction of current flow. Two additional terminals, located on either side of the Hall layer, form the signal output circuit.

When a moving metallic shutter blocks the magnetic field from reaching the Hall layer or element, the Hall-effect switch produces a voltage signal. When the shutter blade moves and allows the magnetic field to expand and reach the Hall element, the Hall-effect switch does not generate voltage signal.

After leaving the Hall layer, the signal is routed to an amplifier where it is strengthened and inverted so that the signal reads high when it is actually coming in low and vice versa. Once it has been inverted, the signal goes through a Schmitt trigger, where it is turned into a clean square wave signal. After conditioning, the signal is sent to the switching unit.

Computer Controlled Ignition System Operation

Computer-controlled ignition systems control the primary circuit and distribute the firing voltages to the spark plugs.

Spark timing varies continuously to obtain optimum air/fuel combustion. The computer determines the best spark timing based on certain engine operating conditions, such as crankshaft position, engine speed, throttle position, engine coolant temperature, and initial and operating manifold or barometric pressure. Once the computer receives input from these and other sensors, it compares the existing operating conditions to information permanently stored or programmed into its memory. The computer matches the existing conditions to a set of conditions stored in its memory, determines proper timing setting, and sends a signal to the ignition module to fire the plugs.

Fuel Injection

In order for an engine to be efficient, it must receive the correct amount of fuel mixed with the correct amount of air. Providing this is the purpose of the fuel injection and fuel delivery systems.

In electronic fuel injection systems, the engine's fuel needs are measured by intake airflow past a sensor or by intake manifold pressure (vacuum). The airflow or manifold vacuum sensor converts its reading to an electrical signal and sends it to the PCM. Here the signal is processed and the current fuel needs are calculated. The PCM then sends an electrical signal to the fuel injectors. This signal determines the amount of time the injector opens and sprays fuel. This interval is known as the injector pulse width.

The PCM is self-regulating and controls the injectors on the basis of operating performance or parameters rather than on preprogrammed instructions. A PCM with a feedback loop, for example, reads the signals from the oxygen sensor, varies the pulse width of the injectors, and again reads the signals from the oxygen sensor. This is repeated until the injectors are pulsed for just the amount of time needed to get the proper amount of oxygen into the exhaust stream. While this interaction is occurring, the system is operating in closed loop. When conditions, such as starting or wide open throttle, demand that the signals from the oxygen sensor be ignored, the system operates in open loop. During open loop, injector pulse length is controlled by set parameters contained in the PCM's memory.

Many other sensors are used in fuel injection systems. Some of the more commonly used are an oxygen sensor, engine coolant temperature sensor, air charge temperature sensor, throttle position sensor, manifold absolute pressure sensor, mass airflow sensor, knock sensor, exhaust gas recirculation valve position sensor, and vehicle speed sensor.

Types of Fuel Injection

Many of the early EFI systems were Throttle Body Injection (TBI) systems in which the fuel was injected above the throttle plates. Recently a similar system, Central Port Injection (CPI), was introduced. In these systems the injector assembly is located in the lower half of the intake manifold. Engines equipped with TBI have gradually become equipped with Port Fuel Injection (PFI), which has injectors located in the intake ports of the cylinders.

Throttle body injection systems have a throttle body assembly mounted on the intake manifold in the position previously occupied by a carburetor. The throttle body assembly usually contains one or two injectors.

On port fuel injection systems, fuel injectors are mounted at the back of each intake valve. Aside from the differences in injector location and number of injectors, operation of throttle body and port systems is quite similar with regard to fuel and air metering, sensors, and computer operation.

Throttle Body Fuel Injection

The basic TBI assembly consists of two major castings: a throttle body with a valve to control airflow and a fuel body to supply the required fuel. A fuel pressure regulator and fuel injector are integral parts of the fuel body. Also included as part of the assembly is a device to control idle speed and one to provide throttle valve positioning data.

The fuel pressure regulator used on the throttle body assembly is similar to a diaphragm-operated relief valve. Fuel pressure is on one side of the diaphragm and atmospheric pressure is on the other side. The regulator is designed to provide a constant pressure on the fuel injector throughout the range of engine loads and speeds. If regulator pressure is too high, a strong fuel odor is emitted and the engine runs too rich. On the other hand, regulator pressure that is too low results in poor engine performance or detonation can take place, due to the lean mixture.

The fuel injector is solenoid operated and pulsed on and off by the vehicle's engine control computer. When the injector's solenoid is energized, a normally closed ball valve is lifted and fuel under pressure is then injected at the walls of the throttle body bore just above the throttle plate.

A fuel injector has a movable armature in its center, and a pintle with a tapered tip is positioned at the lower end of the armature. A spring pushes the armature and pintle downward so the pintle tip seats in the discharge orifice. The injector coil surrounds the armature, and the two ends of the winding are connected to the terminals on the side of the injector. When the ignition switch is turned on, voltage is supplied to one injector terminal and the other terminal is connected through the computer. Each time the control unit completes the circuit from the injector winding to ground, current flows through the injector coil, and the coil magnetism moves the plunger and pintle upward. Under this condition, the pintle tip is unseated from the injector orifice, and fuel sprays out of this orifice.

Port Fuel Injection

PFI systems use one injector at each cylinder. They are mounted in the intake manifold near the cylinder head where they can inject a fine, atomized fuel mist as close as possible to the intake valve. Fuel lines run to each cylinder from a fuel manifold, usually referred to as a fuel rail. Since each cylinder has its own injector, fuel distribution is exactly equal.

The throttle body in a port fuel injection system controls the amount of air that enters the engine.

Port injectors have their tips located in the manifold where constant changes in vacuum would affect the amount of fuel injected. To compensate for these fluctuations, port injection systems are equipped with fuel pressure regulators that sense manifold vacuum and continually adjust the fuel pressure to maintain a constant pressure drop across the injector tips at all times.

When fuel pressure reaches the setting of the regulator, a diaphragm moves against spring tension, and the regulator valve opens. This action allows fuel to flow through the return line to the fuel tank. The fuel pressure drops slightly when the pressure regulator valve opens, and in response, the spring closes the regulator valve.

A vacuum hose is connected from the intake manifold to the vacuum inlet on the pressure regulator. This hose supplies vacuum to the area where the diaphragm spring is located. The vacuum works with the fuel pressure to move the diaphragm and open the valve. When the engine is running at idle speed, high manifold vacuum is supplied to the regulator. Under this condition, the specified fuel pressure opens the regulator valve. When the engine is running under heavy load and/or wide-open throttle, a very low vacuum is supplied to the regulator. During these times, the vacuum does not help open the regulator valve and a higher fuel pressure is required to open the valve.

The change in fuel pressure allows the fuel to be sprayed into the manifold with the same effect, regardless of the pressure present in the manifold. When there is a high vacuum in the manifold, a very low pressure exists and the pressure difference between the fuel spray and the vacuum is the same as when there is a higher pressure in the manifold (low vacuum) and a higher fuel pressure.

Sequential fuel injection systems control each injector individually and according to the engine's firing order. Each injector is opened just before the intake valve opens, which means the mixture is never static in the intake manifold and adjustments to the mixture can be made almost instantaneously between the firing of one injector and the next.

Fuel Delivery System

The components of a typical gasoline delivery system are fuel tanks, fuel lines, fuel filters, and fuel pumps. Fuel is drawn from the fuel tank by an in-tank or chassis-mounted electric fuel pump. Before it reaches the injectors, the fuel passes through a filter that removes dirt and impurities. A fuel line pressure regulator maintains a constant fuel line pressure that may be as high as 50 psi in some systems. This fuel pressure generates the spraying force needed to inject the fuel. Excess fuel not required by the engine returns to the fuel tank through a fuel return line.

Modern fuel tanks include devices that prevent vapors from leaving the tank. They also have an inlet filler tube and cap. Filler tube caps are non-venting and usually have some type of pressure-vacuum relief valve arrangement. Under normal operating conditions the valve is closed, but whenever pressure or vacuum is more than the calibration of the cap, the valve opens. Once the pressure or vacuum has been relieved, the valve closes.

Some form of liquid/vapor separator is incorporated into most vehicles to stop liquid fuel or bubbles from reaching the vapor storage canister or the engine's crankcase. It can be located inside the tank, on the tank, in fuel vent lines, or near the fuel pump.

Inside the fuel tank there is also a sending unit that includes a pick-up tube and float-operated fuel gauge. The fuel tank pick-up tube is connected to the fuel pump by the fuel line. Some electric fuel pumps are combined with the sending unit.

Fuel Lines and Fittings

Fuel lines can be made of either metal tubing or flexible nylon or synthetic rubber hose. The latter must be nonpermeable so that gasoline and gas vapors cannot evaporate through the hose. Similarly, vapor vent lines must be made of materials that resist attack by fuel vapors.

The fuel lines carry fuel from the fuel tank to the fuel pump, fuel filter, and fuel injection assembly. These lines are usually made of rigid metal, although some sections are constructed of rubber hose to allow for car vibrations.

Sections of fuel line are assembled together by fittings. Some of these fittings are a threaded-type fitting, while others are a quick-release design.

Fuel Filters

A strainer, located in the gasoline tank, is made of a finely woven fabric and is designed to prevent large contaminants from entering the fuel system where they could cause excessive fuel pump wear or plug fuel metering devices.

A fuel filter is connected in the fuel line between the fuel tank and the engine. Many of these filters are mounted under the vehicle and others are mounted in the engine compartment.

Fuel Pumps

An electric fuel pump is basically a small DC electric motor with an impeller mounted on the end of the motor's shaft. A pump cover is mounted over the impeller, and this cover contains inlet and discharge ports. When the armature and impeller rotate, fuel is moved from the tank to the inlet port, and the impeller grooves pick up the fuel and force it around the impeller cover and out the discharge port.

Fuel moves from the discharge port through the inside of the motor and out the check valve and outlet connection, which is connected via the fuel line to the fuel filter and underhood fuel system components. A pressure relief valve near the check valve opens if the fuel supply line is restricted and pump pressure becomes very high. When the relief valve opens, fuel is returned through this valve to the pump inlet. When the engine is shut off, the check valve prevents fuel from draining out of the fuel system into the fuel tank.

Electric fuel pump circuits vary depending on the vehicle make and year. In some circuits, an oil pressure switch is connected in series with the fuel pump. The oil pressure switch is wired in series to prevent engine damage from low oil pressure by turning off the fuel pump, making the engine stall.

Not only are fuel pumps controlled to prevent engine damage, they are also controlled to prevent fires during collisions. Various types of sensors are used that detect impact. When impact is sensed, the sensors open the fuel pump circuit. The PCM, in most vehicles, has ultimate control of the fuel pump and will only activate it when the conditions are safe for the passengers and the vehicle itself.

Emission Control Systems

Emission controls on cars and trucks have one purpose: to reduce the amount of pollutants and environmentally damaging substances released by the vehicles.

There are three main automotive pollutants: hydrocarbons (HC), carbon monoxide (CO), and oxides of nitrogen (NO_x). Particulate emissions are also present in diesel engine exhaust. HC emissions are caused largely by unburned fuel from the combustion chambers. HC emissions can also originate from evaporative sources such as the gasoline tank. CO emissions are a by-product of the combustion process and result from incorrect air/fuel mixtures. NO_x emissions are caused by nitrogen and oxygen uniting at cylinder temperatures above 2,500°F (1,371°C).

One of the most important developments for lowering emission levels has been the availability and use of unleaded gasoline. Removing lead from gasoline eliminates the emission of lead particles in an automobile's exhaust. It also increases spark plug life and avoids the formation of lead deposits in the combustion chambers that tend to increase hydrocarbon emissions. Gasoline without lead also allows for the installation of a catalytic converter.

Three basic types of emission control systems are used in vehicles: evaporative control systems, precombustion, and post-combustion. The evaporative control system is a sealed system. It traps the fuel vapors (HC) that would normally escape from the fuel tank into the air.

Many of the pollution control systems used today prevent emissions from being created in the engine, either during or before the combustion cycle. These are precombustion control systems, and include the PCV and EGR systems. Also included in this group are the many advancements made to precisely control the ignition and fuel systems.

Post-combustion control systems clean up the exhaust gases after the fuel has been burned. Secondary air or air injector systems put fresh air into the exhaust to reduce HC and CO to harmless water vapor and carbon dioxide by chemical (thermal) reaction with oxygen in the air. Catalytic converters help this process by reducing NO_x, HC, and CO.

Evaporative Emission Control Systems

The fuel evaporative emission control system reduces the amount of raw fuel vapors that are emitted into the air from the fuel tank. These vapors must not be allowed to escape from the fuel system to the atmosphere.

A charcoal (carbon) canister is located in or near the engine compartment. Fuel vapors from the gas tank are routed to and absorbed onto the surfaces of the canister's charcoal granules. When the vehicle is restarted, the vacuum draws vapors into the intake manifold to be burned in the engine. Canister purging varies widely with make and model. In some instances a fixed restriction allows constant purging whenever there is manifold vacuum. In others, a staged valve provides purging only at speeds above idle. In most systems, the PCM controls a solenoid and purges the canister during conditions that are acceptable.

PCV Systems

Systems designed to prevent or limit the amount of pollutants produced by an engine are called precombustion emission control devices. Although there are specific systems and engine designs that are classified this way, anything that makes an engine more efficient can be categorized as a precombustion emission control. The PCV system is one of them.

During the last part of the engine's combustion stroke, some unburned fuel and products of combustion leak past the engine's piston rings into the crankcase. This leakage into the engine crankcase is called blowby. Blowby must be removed from the engine before it condenses in the crankcase and reacts with the oil to form sludge. Sludge, if allowed to circulate with engine oil, corrodes and accelerates wear of pistons, piston rings, valves, bearings, and other internal working parts of the engine. Blowby gases must also be removed from the crankcase to prevent premature oil leaks. Because these gases enter the crankcase by the pressure formed during combustion, they pressurize the crankcase. The gases exert pressure on the oil pan gasket and crankshaft seals. If the pressure is not relieved, oil is eventually forced out of these seals.

Combustion gases that enter the crankcase are removed by a positive crankcase ventilation system, which uses engine vacuum to draw fresh air through the crankcase. This fresh air, which dissipates the harmful gases, enters through the air filter or through a separate PCV breather filter located on the inside of the air filter housing.

The positive crankcase ventilation system has two major functions. It prevents the emission of blowby gases from the engine crankcase to the atmosphere. It also scavenges the crankcase of vapors that could dilute the oil and cause it to deteriorate or that could build undesirable pressure in the crankcase.

A PCV valve contains a tapered valve. When the engine is not running, a spring keeps the tapered valve seated against the valve housing. During idle or deceleration, the high intake manifold vacuum moves the tapered valve upward against the spring tension. Under this condition, there is a small opening between the tapered valve and the PCV valve housing. Since the engine is not under heavy load during idle or deceleration operation, blowby gases are minimal and the small PCV valve opening is adequate to move the blowby gases out of the crankcase.

Intake manifold vacuum is lower during part-throttle operation than during idle operation. Under this condition, the spring moves the tapered valve downward to increase the opening between this valve and the PCV valve housing. Since engine load is higher at part-throttle operation than at idle operation, blowby gases are increased. The larger opening between the tapered valve and the PCV valve housing allows all the blowby gases to be drawn into the intake manifold.

When the engine is operating under heavy load conditions with a wide throttle opening, the decrease in intake manifold vacuum allows the spring to move the tapered valve further downward in the PCV valve. This action provides a larger opening between the tapered valve and the

PCV valve housing. Since higher engine load results in more blowby gases, the larger PCV valve opening is necessary to allow these gases to flow through the valve into the intake manifold.

Some engines are equipped with a PCV system that does not use a PCV valve. Instead, the blow-by gases are routed into the intake manifold through a fixed orifice tube. The basic system works the same as if it had a valve, except that the system is regulated only by the vacuum on the orifice. The size of the orifice limits the amount of blowby flow into the intake. The engine's air/fuel system is calibrated for this calibrated air leak. Since the action of the PCV allows unmetered air into the intake, the air/fuel system must be set for this amount of extra air.

Knock Sensor and Knock Sensor Module

Many engines with EFI have a knock sensor, or sensors. The knock sensors may be mounted in the block, cylinder head, or intake manifold. A piezoelectric sensing element is mounted in the knock sensor, and a resistor is connected parallel to this sensing element. When the engine detonates, a vibration occurs in the engine. The piezoelectric sensing element changes this vibration to an analog voltage, and this signal is sent to the knock sensor module.

The knock sensor module changes the analog voltage signal to a digital voltage signal and sends this signal to the PCM. When the PCM receives this signal, it reduces the spark advance to prevent detonation.

EGR Systems

The EGR system dilutes the air/fuel mixture with controlled amounts of exhaust gas. Since exhaust gas does not burn, this reduces the peak combustion temperatures. At lower combustion temperatures, very little of the nitrogen in the air combines with oxygen to form NO_x. Most of the nitrogen is simply carried out with the exhaust gases. For driveability, it is desirable to have the EGR valve opening (and the amount of gas flow) proportional to the throttle opening. Driveability is also improved by shutting off the EGR when the engine is started up cold, at idle, and at full throttle. Since the NO_x control requirements vary on different engines, there are several different systems, with various controls to provide these functions.

Typically the EGR valve is a vacuum-operated, flow control valve. On most systems, it is attached to the intake manifold. A small exhaust crossover passage in the intake manifold admits exhaust gases to the inlet port of the EGR valve. Opening the EGR valve by control vacuum at the diaphragm allows exhaust gases to flow through the valve where it mixes with the air/fuel mixture. The effect is to dilute or lean-out the mixture so that it still burns completely but with a reduction in combustion chamber temperatures.

On some engines, the exhaust gas from the EGR system is distributed through passages in the cylinder heads and distribution plates to each intake port. Since the exhaust gas from the EGR system is distributed equally to each cylinder, smoother engine operation results.

In a port vacuum control system, a vacuum line connects the EGR valve to a port in the throttle body above the throttle plate. When the throttle plate is closed, no vacuum is transmitted. The port is exposed to increasing manifold vacuum as the throttle plate opens. The exhaust gas flow rate depends on manifold vacuum, throttle position, and exhaust gas backpressure.

Depending on the system it is used in, the design of the EGR valve will vary. A positive backpressure EGR valve has a bleed port and valve positioned in the center of the diaphragm. A light spring holds this bleed valve open, and an exhaust passage is connected from the lower end of the tapered valve through the stem to the bleed valve. The area under the diaphragm is vented to the atmosphere. When the engine is running, exhaust pressure is applied to the bleed valve. At low engine speeds, exhaust pressure is not high enough to close the bleed valve. If control vacuum is supplied to the diaphragm chamber, the vacuum is bled off through the bleed port, and the valve remains closed.

As engine and vehicle speed increase, the exhaust pressure also increases. At a preset throttle opening, the exhaust pressure closes the EGR valve bleed port. When control vacuum is supplied to the diaphragm, the diaphragm and valve are lifted upward, and the valve is open. If vacuum from an external source is supplied to a positive backpressure EGR valve with the engine not running, the valve will not open, because the vacuum is bled off through the bleed port.

In a negative backpressure EGR valve, a normally closed bleed port is positioned in the center of the diaphragm. An exhaust passage is connected from the lower end of the tapered valve through the stem to the bleed valve. When the engine is running at lower speeds, each time a cylinder fires and an exhaust valve opens, there is a high-pressure pulse in the exhaust system. However, between these high-pressure pulses, there are low-pressure pulses. As the engine speed increases, more cylinder firings occur in a given time, and the high-pressure pulses become closer together in the exhaust system. At low speed, the negative exhaust pulses are more predominant compared to higher engine speeds.

At lower engine and vehicle speeds, the negative pulses in the exhaust system hold the bleed valve open. When the engine and vehicle speed increase to a preset value, the negative exhaust pressure pulses decrease, and the bleed valve closes. Under this condition, if control vacuum is supplied to the diaphragm chamber, the EGR valve is opened. When vacuum from an external source is supplied to a negative backpressure EGR valve with the engine not running, the bleed port is closed, and the vacuum should open the valve.

A digital EGR valve contains up to three electric solenoids that are operated directly by the PCM. Each solenoid contains a movable plunger with a tapered tip that seats in an orifice. When any solenoid is energized, the plunger is lifted and exhaust gas is allowed to recirculate through the orifice into the intake manifold. The solenoids and orifices are different sizes. The PCM can operate one, two, or three solenoids to supply the amount of exhaust recirculation required to provide optimum control of NO_x emissions.

The linear EGR valve contains a single electric solenoid that is operated by the PCM. A tapered pintle is positioned on the end of the solenoid plunger. When the solenoid is energized, the plunger and tapered valve are lifted, and exhaust gas is allowed to recirculate into the intake manifold. The EGR valve contains an EGR valve position (EVP) sensor, which is a linear potentiometer. The signal from this sensor varies from approximately 1 V with the EGR valve closed to 4.5 V with the valve wide open.

The PCM pulses the EGR solenoid winding on and off with pulse width modulation to provide accurate control of the plunger and EGR flow. The EVP sensor acts as a feedback signal to the PCM to inform the PCM if the commanded valve position was achieved.

Catalytic Converters

Post-combustion emission control devices clean up the exhaust after the fuel has been burned but before the gases exit the vehicle's tailpipe. An excellent example of this is the catalytic converter. A converter is one of the most effective emission control devices on a vehicle for reducing HC, CO, and NO_x.

Another post-combustion system is the secondary air or air injection system. This system forces fresh air into the exhaust stream to reduce HC and CO emissions.

A catalytic converter contains a ceramic element coated with a catalyst. A catalyst is something that causes a chemical reaction without being part of the reaction. A catalytic converter causes a chemical change to take place in the passing exhaust gases. Most of the harmful gases are changed to harmless gases.

Three different materials are used as the catalyst in automotive converters: platinum, palladium, and rhodium. Platinum and palladium are the oxidizing elements of a converter. When HC and CO are exposed to heated surfaces covered with platinum and palladium, a chemical reaction takes place. The HC and CO are combined with oxygen to become H_2O and CO_2. Rhodium is a reducing catalyst. When NO_x is exposed to hot rhodium, oxygen is removed and NO_x becomes just N. The removal of oxygen is called reduction, which is why rhodium is a reducing catalyst.

A catalytic converter that contains all three catalysts and reduces HC, CO, and NO_x is called a three-way converter. Catalytic converters that affect only HC and CO are called oxidizing converters. Three-way converters have the oxidizing catalysts in part of the container and the reducing catalyst in the other. Fresh air is injected by the secondary air system between the two catalysts. This air helps the oxidizing catalyst work by making extra oxygen available. The air from the secondary air system is not always forced into the converter. Instead, it is controlled by the secondary air system. Fresh air added to the exhaust at the wrong time could produce NO_x, something the catalytic converter is trying to destroy.

Air Injection Systems

One of the earliest methods used to reduce the amount of hydrocarbons and carbon monoxide in the exhaust was by forcing fresh air into the exhaust system after combustion. This additional fresh air causes further oxidation and burning of the unburned hydrocarbons and carbon monoxide. Oxygen in the air combines with the HC and CO to continue the burning that reduces the HC and CO concentrations. This allows them to oxidize and produce harmless water vapor and carbon dioxide.

The typical pump-type air injection system includes the following:

- An air pump that produces pressurized air that is sent to the exhaust manifold and to the catalytic converter.
- A vacuum controlled air control or air switching valve that is used to route the air from the pump either to the exhaust manifold or to the catalytic converter.
- A thermal vacuum switch that controls the vacuum to the air control valve. When the coolant is cold, it signals the valve to direct air to the exhaust manifold. Then when the engine warms to normal operating temperature, the thermal vacuum switch signals the air control valve to reroute the air to the converter.
- An air by-pass or diverter valve that diverts, or detours, air during deceleration. Excess air in an exhaust rich with fuel can produce a backfire or explosion in a muffler. Air from the pump is diverted to the atmosphere.
- One-way check valves allow air into the exhaust but prevent exhaust from entering the pump in the event that the drive belt breaks.
- Hoses and nozzles are used to distribute and inject the air from the pump.

Pulse-type air injection systems use the natural exhaust pressure pulses to pull air from the air cleaner into the exhaust manifolds and/or the catalytic converter. A manifold pipe is installed in the exhaust manifold for each engine cylinder. The inner end of these pipes is positioned near the exhaust port. The outer ends of the manifold pipes are connected to a metal container, and a one-way check valve is mounted between the outer end of each pipe and the metal container.

At low engine speeds, each negative pressure pulse in the exhaust manifold moves air from the air cleaner through the metal container, check valve, and manifold pipe into an exhaust port. High-pressure pulses in the exhaust manifold close the one-way check valves and prevent exhaust from entering the system.

If the system is used with a catalytic converter, an air check valve and silencer connects in-line between the air cleaner and the catalytic converter. When pressure in the exhaust system is greater than the pressure in the air cleaner, the reed valve inside the air check valve closes. When the pressure in the exhaust system is less than the pressure in the air cleaner, the reed valve opens and draws air into the catalytic converter.

Many late-model air injection systems are controlled by the PCM. The basic components of this system are like the others, except solenoids are used to divert the airflow. Some of these systems are also equipped with an air pump that is driven by an electric motor.

Computer-Controlled Engine Systems

Because all manufacturers have continually updated, expanded, and improved their computerized control systems, there are hundreds of different domestic and import systems on the road. In a computerized engine control system, emission levels, fuel consumption, driveability, and durability are carefully balanced to achieve maximum results with minimum waste. Some of the things engine control systems are designed to do are:

- Air/fuel ratios are held as closely to 14.7 to 1 as possible, allowing maximum catalytic converter efficiency and minimizing fuel consumption.
- Emission control devices, such as EGR valve, carbon canister, and air pump, are operated at predetermined times to increase efficiency.
- The engine is operated as efficiently as possible when it is cold and is warmed up rapidly, reducing unburned hydrocarbon emissions and engine wear due to raw gas washing oil from the piston rings and getting into the crankcase to form sludge and varnish.
- Ignition timing is advanced as much as possible under all conditions.

- Timing and air/fuel ratios are precisely controlled under all operating conditions.
- Control loop operation enables the engine to make rapid changes to match changes in engine temperature, load, and speed.

In a PCM, RAM is used to store data collected by the sensors, the results of calculations, and other information that is constantly changing during engine operation. Information in volatile RAM is erased when the ignition is turned off or when the power to the computer is disconnected. Non-volatile RAM does not lose its data if its power source is disconnected.

The computer's permanent memory is stored in read only memory (ROM) or programmable read only memory (PROM). Like non-volatile RAM, ROM and PROM are not erased when the power source is disconnected. ROM and PROM are used to store computer control system strategy and look-up tables. PROM normally contains the specific information about the vehicle it is installed in.

System adaptive strategy is a plan, created by engine designers and calibration engineers, for the timing and control of computer-controlled systems. In designing the best strategies, it is necessary to look at all the possible conditions an engine may encounter. It is then determined how the system should respond to these conditions.

The look-up tables (sometimes called maps) contain calibrations and specifications. Look-up tables indicate how an engine should perform. For example, information (a reading of 20 in. Hg) is received from the manifold absolute pressure (MAP) sensor. This information, plus information from the engine speed sensor, is compared to a table for spark advance. This table tells the computer what the spark advance should be for that throttle position and engine speed. The computer then modifies the spark advance.

When making decisions, the PCM is constantly referring to three sources of information: the look-up tables, system strategy, and the input from sensors. By comparing information from these sources, the computer makes informed decisions.

If a computer has adaptive strategy capabilities, it can actually learn from past experience. For example, the normal voltage signals from the TP sensor to the PCM range from 0.6 to 4.5 volts. If a 0.2-volt signal is received, the PCM may regard this signal as the result of a worn TP sensor and assign this lower voltage to the normal low voltage signal. In other words, the PCM will add 0.4 volts to the 0.2 volts it received. All future signals from the various throttle positions will also have 0.4 volt added to the signal. Doing this calculation adjusts for the worn TP sensor and ensures that the engine will operate normally. If the input from a sensor is erratic or considerably out of range, the PCM may totally ignore the input.

When a computer has adaptive learning, a short learning period is necessary after the battery has been disconnected, when a computer has been disconnected or replaced, or when the vehicle is new. During this learning period, the engine may surge, idle fast, or have a loss of power. The average learning period lasts for five miles of driving.

Most adaptive strategies have two parts: Short Term Fuel Trim and Long Term Fuel Trim. Short-term strategies are those immediately enacted by the computer to overcome a change in operation. These changes are temporary. Long-term strategies are based on the feedback about the short-term strategies. These changes are more permanent.

To monitor engine conditions, the computer uses a variety of sensors. All sensors perform the same basic function. They detect a mechanical condition (movement or position), chemical state, or temperature condition and change it into an electrical signal that can be used by the PCM to make decisions. The following represent the most commonly used sensors. It should be noted that these sensors are shared between the systems. For example, MAP sensor input is used to control fuel, ignition, EGR, emission system airflow, air intake, and idle speed systems.

Engine Coolant Temperature (ECT) Sensor

The ECT sensor is very important. Its input is used to regulate many engine functions, such as activating and deactivating the early fuel evaporation, EGR, canister purge, and TCC systems and controlling the open- and closed-loop feedback modes of the system.

The ECT sensor is usually located on the cylinder head or intake manifold. The sensor screws into a water jacket. On most systems, the sensor is a variable resistance thermistor. On older systems, a switch coolant sensor may be used. This type of sensor may be designed to remain closed within a certain temperature range or to open only when the engine is warm.

Engine Position Sensors

Engine position sensors tell the computer the speed of the engine and when the piston in each cylinder reaches top dead center (TDC). This input is used to set ignition timing and fuel injection delivery. Several distinct types of engine position sensors are used, but all communicate with the computer by generating a voltage signal. The sensor does this using a Hall-effect switch or magnetic pulse generator. The engine position sensor may be called a distributor pick-up coil, crankshaft or camshaft position sensor, or a profile ignition pick-up sensor.

EGR Valve Position Sensor

Car manufacturers use a variety of sensors or switches to determine when the EGR valve is open. This information is used to adjust the air/fuel mixture. The exhaust gases introduced by the EGR valve into the intake manifold reduce the available oxygen and thus less fuel is needed in order to maintain low HC levels in the exhaust. Most EGR valve position sensors are linear potentiometers mounted on top of the EGR valve. When the EGR valve opens, the potentiometer stem moves upward and a higher voltage signal is sent to the PCM. OBD-II engines are not equipped with these sensors.

Feedback Pressure EGR Sensor

The feedback pressure EGR sensor used in some vehicles is a pressure-sensing voltage divider (functions as a potentiometer) similar to the ones used as MAP and BARO sensors. It senses exhaust pressure in a chamber just under the EGR valve. This pressure causes the sensor to vary its output voltage signal to the computer. When the EGR valve is closed, the pressure in the sensing chamber is equal to exhaust pressure. When the EGR valve opens, pressure in this chamber drops because of the restricting orifice that lets exhaust into the sensing chamber from the exhaust system. The more the valve opens, the more the pressure drops. The feedback pressure EGR sensor's voltage signal tells the computer how far the EGR valve is open. The computer uses this information to fine tune its control of the electronic vacuum regulator, which controls vacuum to the EGR valve. This information also allows the computer to more accurately control air/fuel ratios and ignition timing.

Intake Air Temperature (IAT) Sensor

Also referred to as the air change temperature sensor, this sensor is a thermistor. Its resistance decreases as manifold air temperature increases and increases as manifold air temperature decreases. The PCM measures the voltage drop across the sensor and uses this input to help calculate fuel delivery. Cold intake air is denser; therefore, a richer air-fuel ratio is required. When the IAT signal indicates colder intake air temperature, the PCM provides a richer air-fuel ratio. The input from this sensor may also be used to control the preheated air and early fuel evaporation systems.

Manifold Absolute Pressure (MAP) Sensor

The function of a MAP sensor is to sense air pressure or vacuum in the intake manifold. The computer uses this input as an indication of engine load to adjust the air/fuel mixture and spark timing. The MAP sensor reads vacuum and pressure through a hose connected to the intake manifold. A pressure-sensitive ceramic or silicon element and electronic circuit in the sensor generates a voltage signal that changes in direct proportion to pressure.

MAP sensors should not be confused with vacuum sensors or barometric pressure sensors. While a vacuum sensor reads the difference between manifold vacuum and atmospheric pressure, a MAP sensor measures manifold air pressure against a precalibrated absolute pressure. Because it bases its readings on preset absolute pressure, MAP sensor readings are not adversely altered by changes in operating altitudes or barometric pressure.

Mass Airflow (MAF) Sensor

This sensor measures the flow of air entering the engine. This measurement of airflow is a reflection of engine load (throttle opening and air volume).

It is similar to the relationship of engine load to MAP or vacuum sensor signal. Since there are several types of MAFs, check the service manual for the one used.

In a vane-type MAF, a pivoted air measuring plate is lightly spring-loaded in the closed position. As intake air flows through the sensor, the air measuring plate moves toward the open position. The movement of the plate is proportional to intake airflow. A pointer is attached to the measuring plate shaft. This pointer contacts a resistor to form a potentiometer that sends a voltage signal to the PCM in relationship to intake airflow.

In the heated resistor-type, a heated resistor is mounted in the center of the air passage. When the ignition switch is turned on, voltage is applied to the sensor's module and the module allows enough current flow through the resistor to maintain a specific resistor temperature. If a cold engine is accelerated suddenly, the rush of cool air tries to cool the resistor. Under this condition, more current is needed to maintain the temperature of the resistor. The sensor's module sends a signal to the PCM indicating the amount of current needed to maintain the temperature. When the PCM receives a signal indicating more airflow or higher current, it allows for more fuel to be delivered to the cylinders. Some heated resistor-type MAFs use a heated grid instead of a resistor.

In a hot-wire-type MAF, a hot wire is positioned in the air stream through the sensor, and an ambient temperature sensor wire is located beside the hot wire. This ambient sensor wire, sometimes called the cold wire, senses intake air temperature. When the ignition switch is turned on, the module in the MAF sensor allows enough current flow through the hot wire to allow it to maintain a specific number of degrees above the ambient temperature measured by the cold wire. Like the heated resistor-type MAF, if the engine is suddenly accelerated, the rush of cold air tends to cool the hot wire. As a result, more current is needed to maintain the desired temperature and a signal is sent to the PCM indicating how much current is required. Based on this input, the PCM adds or subtracts fuel in an attempt to achieve the correct air-fuel ratio for the conditions.

Oxygen Sensors (O_2S)

The exhaust gas oxygen sensor is the key sensor in the closed-loop mode. Its input is used by the computer to maintain a balanced air/fuel mixture. The O_2S is threaded into the exhaust manifold or into the exhaust pipe near the engine.

One type of oxygen sensor, made with a zirconium dioxide element, generates a voltage signal proportional to the amount of oxygen in the exhaust gas. It compares the oxygen content in the exhaust gas with the oxygen content of the outside air. As the amount of unburned oxygen in the exhaust gas increases, the voltage output of the sensor drops. Sensor output ranges from 0.1 volt (lean) to 0.9 volt (rich). A perfectly balanced air/fuel mixture of 14.7:1 produces an output of around 0.5 volt. When the sensor reading is lean, the computer enriches the air/fuel mixture to the engine. When the sensor reading is rich, the computer leans the air/fuel mixture.

Because the oxygen sensor must be hot to operate, most late-model engines use heated oxygen sensors (HO_2S). These sensors have an internal heating element that allows the sensor to reach operating temperature more quickly and to maintain its temperature during periods of idling or low engine load.

A second type of oxygen sensor, a titania-type, does not generate a voltage signal. Instead it acts like a variable resistor, altering a base voltage supplied by the control module. When the air/fuel mixture is rich, sensor resistance is low. When the mixture is lean, resistance increases. Variable-resistance oxygen sensors do not need an outside air reference. This eliminates the need for internal venting to the outside. They feature very fast warm-up, and they operate at lower exhaust temperatures.

Throttle Position (TP) Sensor

Engines with electronic fuel injection or feedback carburetors use a TP sensor to inform the computer about the rate of throttle opening and the relative throttle position. The TP sensor contains a potentiometer with a pointer that is rotated by the throttle shaft. As the throttle shaft moves, the pointer moves to a new location on the resistor in the potentiometer. The return voltage signal tells the PCM how much the throttle plates are open. As resistance readings tell the computer that the throttle is opening, it enriches the air/fuel mixture to maintain the proper air-fuel ratio. A separate idle switch or wide open throttle (WOT) switch may also be used to signal the computer when these throttle positions exist.

The initial setting of the sensor is critical. The voltage signal the computer receives is referenced to this setting. Many service manuals list the initial TP sensor setting to the nearest one-hundredth volt, a clear indication of the importance of this setting.

Vehicle Speed Sensor (VSS)

This sensor tells the computer the vehicle's speed in miles per hour. This input controls when the torque converter clutch locks up and also can be used to control cruise control, EGR flow, and canister purge. The VSS is connected in the speedometer cable or mounted in the transmission/transaxle opening where the speedometer drive was located previously.

Computer Outputs and Actuators

Once the PCM's programming instructs that a correction or adjustment must be made in the controlled system, an output signal is sent to a control device or actuator. These actuators, which are solenoids, switches, relays, or motors, physically act or carry out the command sent by the PCM.

Actuators are electromechanical devices that convert an electrical current into mechanical action. This mechanical action can then be used to open and close valves, control vacuum to other components, or open and close switches. When the PCM receives an input signal indicating a change in one or more of the operating conditions, the PCM determines the best strategy for handling the conditions. The PCM then controls a set of actuators to achieve a desired effect or strategy goal. In order for the computer to control an actuator, it must rely on an output driver.

The circuit driver usually applies the ground circuit of the actuator. The ground can be applied steadily if the actuator must be activated for a selected amount of time. Or the ground can be pulsed to activate the actuator in pulses. Output drivers are transistors or groups of transistors that control the actuators. These drivers operate by the digital commands from the PCM. If an actuator can't be controlled digitally, the output signal must pass through an A/D converter before flowing to the actuator. The major actuators in a computer-controlled engine include the following components.

- *Air Management Solenoids*—Secondary air by-pass and diverter solenoids control the flow of air from the air pump to either the exhaust manifold (open loop) or the catalytic converter (closed loop).

- *Evaporative Emission (EVAP) Canister Purge Valve*—This valve is controlled by a solenoid. The valve controls when stored fuel vapors in the canister are drawn into the engine and burned. The computer only activates this solenoid valve when the engine is warm and above idle speed.

- *EGR Flow Solenoids*—EGR flow may be controlled by electronically controlled vacuum solenoids. The solenoid valves supply manifold vacuum to the EGR valve when EGR is required or may vent vacuum when EGR is not required.

- *Fuel Injectors*—These solenoid valves deliver the fuel spray in fuel-injected systems.

- *Ignition Module*—This is actually an electronic switching device triggered by a signal from the control computer. The ignition module may be a separate unit or may be part of the PCM.

- *Motors and Lights*—Using electrical relays, the computer is used to trigger the operation of electric motors such as the fuel pump, or various warning light or display circuits.

- *Other Solenoids*—Computer-controlled solenoids may also be used in the operation of cruise control systems, torque converter clutches, automatic transmission shift mechanisms, and many other systems where mechanical action is needed.

System Operation

Control loops are the cycles by which a process can be controlled by information received from input sensors, ROM, computer processing, and output of specific commands to control actuator devices.

The basic purpose of all computerized engine control loops is the same: to create an ideal air-fuel ratio, which allows the catalytic converter to operate at maximum efficiency, while giving the best mileage and performance possible and protecting the engine.

The closed loop mode is basically the same for any automotive system. Sensor inputs are sent to the computer, the computer compares the values to its programs, and then sends commands to the output devices. The output devices adjust timing, air-fuel ratio, and emission control operation. The resulting engine operation affects the sensors, which send new messages to the computer, completing the cycle of operation. The complete cycle is called a closed loop.

Closed control loops are often referred to as feedback systems. This means that the sensors provide constant information, or feedback, on what is taking place in the engine. This allows the computer to make constant decisions and changes to output commands.

When the engine is cold, most electronic engine controls go into open loop mode. In this mode, the control loop is not a complete cycle because the computer does not react to feedback information. Instead, the computer makes decisions based on preprogrammed information that allows it to make basic ignition or air/fuel settings and to disregard sensor inputs. The open loop mode is activated when a signal from the temperature sensor indicates that the engine temperature is too low for gasoline to properly vaporize and burn in the cylinders. Systems with oxygen sensors may also go into the open loop mode while idling, or at any time that the oxygen sensor cools off enough to stop sending a signal, and at wide open throttle.

SAFETY

In an automotive repair shop, there is great potential for serious accidents, simply because of the nature of the business and the equipment used. When people are careless, the automotive repair industry can be one of the most dangerous occupations. But the chances of your being injured while working on a car are close to nil if you learn to work safely and use common sense. Safety is the responsibility of everyone in the shop.

Personal Protection

Some procedures, such as grinding, result in tiny particles of metal and dust that are thrown off at very high speeds. These metal and dirt particles can easily get into your eyes, causing scratches or cuts on your eyeball. Pressurized gases and liquids escaping a ruptured hose or hose-fitting can spray a great distance. If these chemicals get into your eyes, they can cause blindness. Dirt and sharp bits of corroded metal can easily fall down into your eyes while you are working under a vehicle.

Eye protection should be worn whenever you are exposed to these risks. To be safe, you should wear safety glasses whenever you are working in the shop. Some procedures may require that you wear other eye protection in addition to safety glasses. For example, when cleaning parts with a pressurized spray, you should wear a face shield. The face shield not only gives added protection to your eyes but also protects the rest of your face.

If chemicals such as battery acid, fuel, or solvents get into your eyes, flush them continuously with clean water. Have someone call a doctor and get medical help immediately.

Your clothing should be well fitted and comfortable but made of strong material. Loose, baggy clothing can easily get caught in moving parts and machinery. Some technicians prefer to wear coveralls or shop coats to protect their personal clothing. Your work clothing should offer you some protection but should not restrict your movement.

Long hair and loose, hanging jewelry can create the same type of hazard as loose-fitting clothing. They can get caught in moving engine parts and machinery. If you have long hair, tie it back or tuck it under a cap.

Never wear rings, watches, bracelets, and neck chains. These can easily get caught in moving parts and cause serious injury.

Always wear shoes or boots of leather or similar material with non-slip soles. Steel-tipped safety shoes can give added protection to your feet. Jogging or basketball shoes, street shoes, and sandals are inappropriate in the shop.

Good hand protection is often overlooked. A scrape, cut, or burn can limit your effectiveness at work for many days. A well-fitted pair of heavy work gloves should be worn during operations such as grinding and welding or when handling hot components. Always wear approved rubber gloves when handling strong and dangerous caustic chemicals.

Many technicians wear thin, surgical-type latex gloves whenever they are working on vehicles. These offer little protection against cuts but do offer protection against disease and grease buildup under and around your fingernails. These gloves are comfortable and are quite inexpensive.

Accidents can be prevented simply by the way you act. The following are some guidelines to follow while working in a shop. This list does not include everything you should or shouldn't do; it merely presents some things for you to think about.

- Never smoke while working on a vehicle or while working with any machine in the shop.
- Playing around is not fun when it sends someone to the hospital.
- To prevent serious burns, keep your skin away from hot metal parts such as the radiator, exhaust manifold, tailpipe, catalytic converter, and muffler.
- Always disconnect electric engine cooling fans when working around the radiator. Many of these will turn on without warning and can easily chop off a finger or hand. Make sure you reconnect the fan after you have completed your repairs.
- When working with a hydraulic press, make sure the pressure is applied in a safe manner. It is generally wise to stand to the side when operating the press.
- Properly store all parts and tools by putting them away in a place where people will not trip over them. This practice not only cuts down on injuries, but also reduces time wasted looking for a misplaced part or tool.

Work Area Safety

Your entire work area should be kept clean and safe. Any oil, coolant, or grease on the floor can make it slippery. To clean up oil, use commercial oil absorbent. Keep all water off the floor. Water not only makes smooth floors slippery, but it is also dangerous as a conductor of electricity. Aisles and walkways should be kept clean and wide enough to allow easy movement. Make sure the work areas around machines are large enough so that the machines can be safely operated.

Gasoline is a highly flammable volatile liquid. Something that is flammable catches fire and burns easily. A volatile liquid is one that vaporizes very quickly. Flammable volatile liquids are potential firebombs. Always keep gasoline or diesel fuel in an approved safety can and never use gasoline to clean your hands or tools.

Handle all solvents (or any liquids) with care to avoid spillage. Keep all solvent containers closed, except when pouring. Proper ventilation is very important in areas where volatile solvents and chemicals are used. Solvents and other combustible materials must be stored in approved and designated storage cabinets or rooms with adequate ventilation. Never light matches or smoke near flammable solvents and chemicals, including battery acids.

Oily rags should also be stored in an approved metal container. When these oily, greasy, or paint-soaked rags are left lying about or are not stored properly, they can cause spontaneous combustion. Spontaneous combustion results in a fire that starts by itself, without a match.

Disconnecting the vehicle's battery before working on the electrical system, or before welding, can prevent fires caused by a vehicle's electrical system. To disconnect the battery, remove the negative or ground cable from the battery and position it away from the battery.

Know where all of the shop's fire extinguishers are located. Fire extinguishers are clearly labeled as to what type they are and what types of fire they should be used on. Make sure you use the correct type of extinguisher for the type of fire you are dealing with. A multipurpose dry chemical fire extinguisher will put out ordinary combustibles, flammable liquids, and electrical fires. Never put water on a gasoline fire because that will just cause the fire to spread. The proper fire extinguisher will smother the flames.

During a fire, never open doors or windows unless it is absolutely necessary; the extra draft will only make the fire worse. Make sure the fire department is contacted before or during your attempt to extinguish a fire.

Battery Safety

The potential dangers caused by the sulfuric acid in the electrolyte and the explosive gases generated during battery charging require that battery service and troubleshooting be conducted under absolutely safe working conditions. Always wear safety glasses or goggles when working with batteries no matter how small the job.

Sulfuric acid can also cause severe skin burns. If electrolyte contacts your skin or eyes, flush the area with water for several minutes. When eye contact occurs, force your eyelid open. Always have a bottle of neutralizing eyewash on hand and flush the affected areas with it. Do not rub your eyes or skin. Receive prompt medical attention if electrolyte contacts your skin or eyes. Call a doctor immediately.

When a battery is charging or discharging, it gives off quantities of highly explosive hydrogen gas. Some hydrogen gas is present in the battery at all times. Any flame or spark can ignite this gas, causing the battery to explode violently, propelling the vent caps at a high velocity and spraying acid over a wide area. To prevent this dangerous situation, take these precautions:

- Never smoke near the top of a battery and never use a lighter or match as a flashlight.
- Remove wristwatches and rings before servicing any part of the electrical system. This helps to prevent the possibility of electrical arcing and burns.
- Even sealed, maintenance-free batteries have vents and can produce dangerous quantities of hydrogen if severely overcharged.
- When removing a battery from a vehicle, always disconnect the battery ground cable first. When installing a battery, connect the ground cable last.
- Always disconnect the battery's ground cable when working on the electrical system or engine. This prevents sparks from short circuits and prevents accidental starting of the engine.
- Always operate charging equipment in well-ventilated areas. A battery that has been overworked should be allowed to cool down. Let air circulate around it before attempting to jump-start the vehicle. Most batteries have flame arresters in the caps to help prevent explosions, so make sure that the caps are tightly in place.
- Never connect or disconnect charger leads when the charger is turned on. This generates a dangerous spark.
- Never lay metal tools or other objects on the battery because a short circuit across the terminals can result.
- Always disconnect the battery ground cable before fast-charging the battery on the vehicle. Improper connection of charger cables to the battery can reverse the current flow and damage the AC generator.
- Never attempt to use a fast charger as a boost to start the engine.
- As a battery gets closer to being fully discharged, the acidity of the electrolyte is reduced, and the electrolyte starts to behave more like pure water. A dead battery may freeze at temperatures near zero degrees Fahrenheit. Never try to charge a battery that has ice in the cells. Passing current through a frozen battery can cause it to rupture or explode. If ice or slush is visible or the electrolyte level cannot be seen, allow the battery to thaw at room temperature before servicing. Do not take chances with sealed batteries. If there is any doubt, allow them to warm to room temperature before servicing.
- As batteries get old, especially in warm climates and especially with lead-calcium cells, the grids start to grow. The chemistry is rather involved, but the point is that plates can grow to the point where they touch, producing a shorted cell.
- Always use a battery carrier or lifting strap to make moving and handling batteries easier and safer.
- Acid from the battery damages a vehicle's paint and metal surfaces and harms shop equipment. Neutralize any electrolyte spills during servicing.

Air Bag Safety

When service is performed on any air bag system component, always disconnect the negative battery cable, isolate the cable end, and wait for the amount of time specified by the vehicle manufacturer before proceeding with the necessary diagnosis or service. The average waiting period is two minutes, but some vehicle manufacturers specify up to ten minutes. Failure to observe this precaution may cause accidental air bag deployment and personal injury.

Replacement air bag system parts must have the same part number as the original part. Replacement parts of lesser or questionable quality must not be used. Improper or inferior components may result in inappropriate air bag deployment and injury to vehicle occupants.

Do not strike or jar a sensor or an air bag system diagnostic monitor (ASDM). This may cause air bag deployment or make the sensor inoperative. Accidental air bag deployment may cause personal injury, and an inoperative sensor may result in air bag deployment failure, causing personal injury to vehicle occupants.

All sensors and mounting brackets must be properly torqued to ensure correct sensor operation before an air bag system is powered up. If sensor fasteners do not have the proper torque, improper air bag deployment may result in injury to vehicle occupants.

When working on the electrical system on an air-bag-equipped vehicle, use only the vehicle manufacturer's recommended tools and service procedures. The use of improper tools or service procedures may cause accidental air bag deployment and personal injury. For example, do not use 12V or self-powered test lights when servicing the electrical system on an air-bag-equipped vehicle.

Tool and Equipment Safety

Careless use of simple hand tools such as wrenches, screwdrivers, and hammers causes many shop accidents that could be prevented. Keep all hand tools grease-free and in good condition. Tools that slip can cause cuts and bruises. If a tool slips and falls into a moving part, it can fly out and cause serious injury.

Use the proper tool for the job. Make sure the tool is of professional quality. Using poorly made tools or the wrong tools can damage parts or the tool itself, or could cause injury. Never use broken or damaged tools.

Safety around power tools is very important. Serious injury can result from carelessness. Always wear safety glasses when using power tools. If the tool is electrically powered, make sure it is properly grounded. Before using it, check the wiring for cracks in the insulation, as well as for bare wires. Also, when using electrical power tools, never stand on a wet or damp floor. Never leave a running power tool unattended.

When using compressed air, safety glasses and/or a face shield should be worn. Particles of dirt and pieces of metal, blown by the high-pressure air, can penetrate your skin or get into your eyes.

Always be careful when raising a vehicle on a lift or a hoist. Adapters and hoist plates must be positioned correctly to prevent damage to the underbody of the vehicle. There are specific lift points that allow the weight of the vehicle to be evenly supported by the adapters or hoist plates. The correct lift points can be found in the vehicle's service manual. Before operating any lift or hoist, carefully read the operating manual and follow the operating instructions.

Once you feel the lift supports are properly positioned under the vehicle, raise the lift until the supports contact the vehicle. Then, check the supports to make sure they are in full contact with the vehicle. Shake the vehicle to make sure it is securely balanced on the lift, and then raise the lift to the desired working height. Before working under a car, make sure the lift's locking devices are engaged.

A vehicle can be raised off the ground by a hydraulic jack. The jack's lifting pad must be positioned under an area of the vehicle's frame or at one of the manufacturer's recommended lift points. Never place the pad under the floor pan or under steering and suspension components, because these are easily damaged by the weight of the vehicle. Always position the jack so the wheels of the vehicle can roll as the vehicle is being raised.

Safety stands, also called jack stands, should be placed under a sturdy chassis member, such as the frame or axle housing, to support the vehicle after it has been raised by a jack. Once the safety stands are in position, the hydraulic pressure in the jack should be slowly released until the weight of the vehicle is on the stands. Never move under a vehicle when it is supported only by a hydraulic jack. Rest the vehicle on the safety stands before moving under the vehicle.

Parts cleaning is a necessary step in most repair procedures. Always wear the appropriate protection when using chemical, abrasive, and thermal cleaners.

Vehicle Operation

When the customer brings a vehicle in for service, certain driving rules should be followed to ensure your safety and the safety of those working around you. For example, before moving a car into the shop, buckle your safety belt. Make sure no one is near, the way is clear, and there are no tools or parts under the car before you start the engine. Check the brakes before putting the vehicle in gear. Then, drive slowly and carefully in and around the shop.

If the engine must be running while you are working on the car, block the wheels to prevent the car from moving. Place the transmission into park for automatic transmissions or into neutral for manual transmissions. Set the parking (emergency) brake. Never stand directly in front of or behind a running vehicle.

Run the engine only in a well-ventilated area to avoid the danger of poisonous carbon monoxide (CO) in the engine exhaust. CO is an odorless but deadly gas. Most shops have an exhaust ventilation system, and you should always use it. Connect the hose from the vehicle's tailpipe to the intake for the vent system. Make sure the vent system is turned on before running the engine. If the work area does not have an exhaust venting system, use a hose to direct the exhaust out of the building.

HAZARDOUS MATERIALS AND WASTES

A typical shop contains many potential health hazards for those working in it. These hazards can cause injury, sickness, health impairments, discomfort, and even death. Here is a short list of the different classes of hazards:

- Chemical hazards are caused by high concentrations of vapors, gases, or solids in the form of dust.
- Hazardous wastes are those substances that result from a service being performed.
- Physical hazards include excessive noise, vibration, pressure, and temperature.
- Ergonomic hazards are conditions that impede normal and/or proper body position and motion.

There are many government agencies charged with ensuring safe work environments for all workers. These include the Occupational Safety and Health Administration (OSHA), Mine Safety and Health Administration (MSHA), and National Institute for Occupational Safety and Health (NIOSH). These, as well as state and local governments, have instituted regulations that must be understood and followed. Everyone in a shop has the responsibility for adhering to these regulations.

An important part of a safe work environment is the employees' knowledge of potential hazards. Right-to-know laws concerning all chemicals protect every employee in the shop. The general intent of right-to-know laws is to ensure that employers provide their employees with a safe working place as far as hazardous materials are concerned.

All employees must be trained about their rights under the legislation, the nature of the hazardous chemicals in their workplace, and the contents of the labels on the chemicals. All of the information about each chemical must be posted on material safety data sheets (MSDS) and must be accessible. The manufacturer of the chemical must give these sheets to its customers, if they are requested to do so. The sheets detail the chemical composition and precautionary information for all products that can present a health or safety hazard.

Employees must become familiar with the general uses, protective equipment, accident or spill procedures, and any other information regarding the safe handling of a particular hazardous material. This training must be given to employees annually and provided to new employees as part of their job orientation.

All hazardous material must be properly labeled, indicating what health, fire, or reactivity hazard it poses and what protective equipment is necessary when handling each chemical. The manufacturer of the hazardous materials must provide all warnings and precautionary information, which must be read and understood by the user before use. A list of all hazardous materials used in the shop must be posted for the employees to see.

Shops must maintain documentation on the hazardous chemicals in the workplace, proof of training programs, records of accidents or spill incidents, satisfaction of employee requests for specific chemical information via the MSDS, and a general right-to-know compliance procedure manual utilized within the shop.

When handling any hazardous materials or hazardous waste, make sure you follow the required procedures for handling such material. Also wear the proper safety equipment listed on the MSDS. This includes the use of approved respirator equipment.

Some of the common hazardous materials that automotive technicians use are: cleaning chemicals, fuels (gasoline and diesel), paints and thinners, battery electrolyte (acid), used engine oil, refrigerants, and engine coolant (anti-freeze).

Many repair and service procedures generate what are known as hazardous wastes. Dirty solvents and cleaners are good examples of hazardous wastes. Something is classified as a hazardous waste if it is on the EPA list of known harmful materials or has one or more of the following characteristics.

- *Ignitability.* If it is a liquid with a flash point below 140°F or a solid that can spontaneously ignite.
- *Corrosivity.* If it dissolves metals and other materials or burns the skin.
- *Reactivity.* Any material that reacts violently with water or other materials or releases cyanide gas, hydrogen sulfide gas, or similar gases when exposed to low pH acid solutions. This also includes material that generates toxic mists, fumes, vapors, and flammable gases.
- *Toxicity.* Materials that leach one or more of eight heavy metals in concentrations greater than 100 times primary drinking water standard concentrations.

Complete EPA lists of hazardous wastes can be found in the Code of Federal Regulations. It should be noted that no material is considered hazardous waste until the shop is finished using it and ready to dispose of it.

The following list covers the recommended procedure for dealing with some of the common hazardous wastes. Always follow these and any other mandated procedures.

Oil Recycle oil. Set up equipment, such as a drip table or screen table with a used oil collection bucket, to collect oils dripping off parts. Place drip pans underneath vehicles that are leaking fluids onto the storage area. Do not mix other wastes with used oil, except as allowed by your recycler. Used oil generated by a shop (and/or oil received from household "do-it-yourself" generators) may be burned on site in a commercial space heater. Also, used oil may be burned for energy recovery. Contact state and local authorities to determine requirements and to obtain the necessary permits.

Oil filters Drain for at least 24 hours, crush, and recycle used oil filters.

Batteries Recycle batteries by sending them to a reclaimer or back to the distributor. Keeping shipping receipts can demonstrate that you have done the recycling. Store batteries in a watertight, acid-resistant container. Inspect batteries for cracks and leaks when they come in. Treat a dropped battery as if it were cracked. Acid residue is hazardous because it is corrosive and may contain lead and other toxic substances. Neutralize spilled acid, by using baking soda or lime, and dispose of it as hazardous material.

Metal residue from machining Collect metal filings when machining metal parts. Keep them separate and recycle if possible. Prevent metal filings from falling into a storm sewer drain.

Refrigerants Recover and/or recycle refrigerants during the servicing and disposal of motor vehicle air conditioners and refrigeration equipment. It is not allowable to knowingly vent refrigerants to the atmosphere. Recovering and/or recycling during servicing must be performed by an EPA-certified technician using certified equipment and following specified procedures.

Solvents Replace hazardous chemicals with less toxic alternatives that perform equally. For example, substitute water-based cleaning solvents for petroleum-based solvent degreasers. To reduce the amount of solvent used when cleaning parts, use a two-stage process: dirty solvent followed by fresh solvent. Hire a hazardous waste management service to clean and recycle solvents. (Some spent solvents must be disposed of as hazardous waste, unless recycled properly). Store solvents in closed containers to prevent evaporation. Evaporation of solvents contributes to ozone depletion and smog formation. In addition, the residue from evaporation must be treated as a hazardous waste. Properly label spent solvents and store them on drip pans or in diked areas and only with compatible materials.

Containers Cap, label, cover, and properly store, aboveground outdoor liquid containers and small tanks within a diked area and on a paved impermeable surface to prevent spills from running into surface or ground water.

Other solids Store materials such as scrap metal, old machine parts, and worn tires under a roof or tarpaulin to protect then from the elements and to prevent the possibility of creating contaminated runoff. Consider recycling tires by retreading them.

Liquid recycling Collect and recycle coolants from radiators. Store transmission fluids, brake fluids, and solvents containing chlorinated hydrocarbons separately, and recycle or dispose of them properly.

Shop towels and rags Keep waste towels in a closed container marked "contaminated shop towels only." To reduce costs and liabilities associated with disposal of used towels, which can be classified as hazardous wastes, investigate using a laundry service that is able to treat the wastewater generated from cleaning the towels.

Waste storage Always keep hazardous waste separate, properly labeled, and sealed in the recommended containers. The storage area should be covered and may need to be fenced and locked if vandalism could be a problem. Select a licensed hazardous waste hauler after seeking recommendations and reviewing the firm's permits and authorizations.

NATEF TASK LIST FOR ENGINE PERFORMANCE

A. General Engine Diagnosis

A.1. Complete work order to include customer information, vehicle identifying information, customer concern, related service history, cause, and correction. Priority Rating 1

A.2. Identify and interpret engine performance concern; determine necessary action. Priority Rating 1

A.3. Research applicable vehicle and service information, such as engine management system operation, vehicle service history, service precautions, and technical service bulletins. Priority Rating 1

A.4. Locate and interpret vehicle and major component identification numbers (VIN, vehicle certification labels, calibration decals). Priority Rating 1

A.5. Inspect engine assembly for fuel, oil, coolant, and other leaks; determine necessary action. Priority Rating 2

A.6. Diagnose unusual engine noise or vibration concerns; determine necessary action. Priority Rating 2

A.7. Diagnose unusual exhaust color, odor, and sound; determine necessary action. Priority Rating 2

A.8. Perform engine absolute (vacuum/boost) manifold pressure tests; determine necessary action. Priority Rating 1

A.9. Perform cylinder power balance test; determine necessary action. Priority Rating 1

A.10. Perform cylinder cranking compression test; determine necessary action. Priority Rating 1

A.11. Perform engine running compression test; determine necessary action. Priority Rating 2

A.12. Perform cylinder leakage test; determine necessary action. Priority Rating 1

A.13. Diagnose engine mechanical, electrical, electronic, fuel, and ignition concerns with an oscilloscope and engine diagnostic equipment; determine necessary action. Priority Rating 1

A.14. Prepare 4 or 5 gas analyzer; inspect and prepare vehicle for test, and obtain exhaust readings; interpret readings, and determine necessary action. Priority Rating 1

A.15. Verify engine operating temperature; determine necessary action. Priority Rating 1

A.16. Perform engine cooling system pressure tests; check coolant condition; inspect and test radiator, pressure cap, coolant recovery tank, and hoses; perform necessary action. Priority Rating 1

A.17. Verify correct camshaft timing. Priority Rating 2

B. Computerized Engine Controls Diagnosis and Repair

B.1. Retrieve and record stored OBD I diagnostic trouble codes; clear codes. Priority Rating 3

B.2. Retrieve and record stored OBD II diagnostic trouble codes; clear codes, when applicable. Priority Rating 1

B.3. Diagnose the causes of emissions or driveability concerns resulting from malfunctions in the computerized engine control system with stored diagnostic trouble codes. Priority Rating 1

B.4. Diagnose emissions or driveability concerns resulting from malfunctions in the computerized engine controls with no stored diagnostic trouble codes; determine necessary action. Priority Rating 1

B.5. Check for module communication (including CAN/BUS systems) errors using a scan tool. Priority Rating 1

B.6. Inspect and test computerized engine control system sensors, power train control module (PCM), actuators, and circuits using a graphing multimeter (GMM)/digital storage oscilloscope (DSO); perform necessary action. Priority Rating 1

B.7. Obtain and interpret scan tool data. Priority Rating 1

B.8. Access and use service information to perform step-by-step diagnosis. Priority Rating 1

B.9. Diagnose driveability and emissions problems resulting from malfunctions of interrelated systems (cruise control, security alarms, suspension controls, traction controls, A/C, automatic transmissions, non-OEM-installed accessories, and similar systems); determine necessary action. Priority Rating 3

B.10. Perform active tests of actuators using scan tool; determine necessary action. Priority Rating 1

C. Ignition System Diagnosis and Repair

C.1. Diagnose ignition system related problems such as no-starting, hard starting, engine misfire, poor driveability, spark knock, power loss, poor mileage, and emissions concerns on vehicles with electronic ignition (distributorless) systems; determine necessary action. Priority Rating 1

C.2. Diagnose ignition system related problems such as no-starting, hard starting, engine misfire, poor driveability, spark knock, power loss, poor mileage, and emissions concerns on vehicles with distributor ignition (DI) systems; determine necessary action. Priority Rating 1

C.3. Inspect and test ignition primary circuit wiring and solid state components; perform necessary action. Priority Rating 2

C.4. Inspect, test, and service distributor. Priority Rating 3

C.5. Inspect and test ignition system secondary circuit wiring and components; perform necessary action. Priority Rating 2

C.6. Inspect and test ignition coil(s); perform necessary action. Priority Rating 1

C.7. Check and adjust ignition system timing and timing advance/retard (where applicable). Priority Rating 3

C.8. Inspect and test ignition system pick-up sensor or triggering devices; perform necessary action. Priority Rating 3

D. Fuel, Air Induction, and Exhaust Systems Diagnosis and Repair

D.1. Diagnose hot or cold no-starting, hard starting, poor driveability, incorrect idle speed, poor idle, flooding, hesitation, surging, engine misfire, power loss, stalling, poor mileage, dieseling, and emissions problems on vehicles with injection-type fuel systems; determine necessary action. Priority Rating 1

D.2. Check fuel for contaminants and quality; determine necessary action. Priority Rating 3

D.3. Inspect and test fuel pumps and pump control systems for pressure, regulation, and volume; perform necessary action. Priority Rating 1

D.4. Replace fuel filters. Priority Rating 1

D.5. Inspect and test cold enrichment system and components; perform necessary action. Priority Rating 3

D.6. Inspect throttle body, air induction system, intake manifold and gaskets for vacuum leaks and/or unmetered air. Priority Rating 2

D.7. Inspect and test fuel injectors. Priority Rating 1

D.8. Check idle speed. Priority Rating 2

D.9. Inspect the integrity of the exhaust manifold, exhaust pipes, muffler(s), catalytic converter(s), resonator(s), tail pipe(s), and heat shield(s); perform necessary action. Priority Rating 2

D.10. Perform exhaust system back-pressure test; determine necessary action. Priority Rating 1

D.11. Test the operation of turbocharger/supercharger systems; determine necessary action. Priority Rating 3

E. Emissions Control Systems Diagnosis and Repair

1. *Positive Crankcase Ventilation*

E.1.1. Diagnose oil leaks, emissions, and driveability problems resulting from malfunctions in the positive crankcase ventilation (PCV) system; determine necessary action. Priority Rating 2

E.1.2. Inspect, test, and service positive crankcase ventilation (PCV) filter/breather cap, valve, tubes, orifices, and hoses; perform necessary action. Priority Rating 2

2. *Exhaust Gas Recirculation*

E.2.1. Diagnose emissions and driveability problems caused by malfunctions in the exhaust gas recirculation (EGR) system; determine necessary action. Priority Rating 1

E.2.2. Inspect, test, service, and replace components of the EGR system, including EGR tubing, exhaust passages, vacuum/pressure controls, filters and hoses; perform necessary action. Priority Rating 1

E.2.3. Inspect and test electrical/electronic sensors, controls, and wiring of exhaust gas recirculation (EGR) systems; perform necessary action. Priority Rating 2

3. *Exhaust Gas Treatment*

E.3.1. Diagnose emissions and driveability problems resulting from malfunctions in the secondary air injection and catalytic converter systems; determine necessary action. Priority Rating 2

E.3.2. Inspect and test mechanical components of secondary air injection systems; perform necessary action. Priority Rating 3

E.3.3. Inspect and test electrical/electronically-operated components and circuits of air injection systems; perform necessary action. Priority Rating 3

E.3.4. Inspect and test catalytic converter performance. Priority Rating 1

4. *Evaporative Emissions Controls*
   - E.4.1. Diagnose emissions and driveability problems resulting from malfunctions in the evaporative emissions control system; determine necessary action. Priority Rating 1
   - E.4.2. Inspect and test components and hoses of evaporative emissions control system; perform necessary action. Priority Rating 2
   - E.4.3. Interpret evaporative emission related diagnostic trouble codes (DTCs); determine necessary action. Priority Rating 1

F. Engine Related Service

- F.1. Adjust valves on engines with mechanical or hydraulic lifters. Priority Rating 1
- F.2. Remove and replace timing belt; verify correct camshaft timing. Priority Rating 1
- F.3. Remove and replace thermostat and gasket. Priority Rating 1
- F.4. Inspect and test mechanical/electrical fans, fan clutch, fan shroud/ducting, air dams, and fan control devices; perform necessary action. Priority Rating 1
- F.5. Perform common fastener and thread repair to include, remove broken bolt, restore internal and external threads, and repair internal threads with thread insert. Priority Rating 1
- F.6. Perform oil and filter changes. Priority Rating 1
- F.7. Demonstrate proficiency in using oxy-acetylene torch to heat and cut metal. Priority Rating 3
- F.8. Identify hybrid vehicle internal combustion engine service precautions. Priority Rating 3

DEFINITION OF TERMS USED IN THE TASK LIST

To clarify the intent of these tasks, NATEF has defined some of the terms used in the task listings. To get a good understanding of what the task includes, refer to this glossary while reading the task list.

adjust To bring components to specified operational settings.

analyze To examine the relationship of components of an operation.

assemble (reassemble) To fit together the components of a device.

bleed To remove air from a closed system.

CAN/LAN/BUS Refers to a high-speed communications system designed to link computers and other devices together. In-vehicle networking involving time division is also known as multiplexing. It is a method for transferring data among electronic modules via a serial data bus. Serial data is electronically coded information which is transmitted by one computer and received and displayed by another computer. The term serial data implies that the information is digitally coded and transmitted in a series of data words.

charge To bring to "full" state, e.g., battery or air conditioning system.

check To verify condition by performing an operational or comparative examination.

clean To rid components of extraneous matter for the purpose of reconditioning, repairing, measuring and reassembling.

determine To establish the procedure to be used to effect the necessary repair.

determine necessary action Indicates that the diagnostic routine(s) is the primary emphasis of a task. The student is required to perform the diagnostic steps and communicate the diagnostic outcomes and corrective actions

	required addressing the concern or problem. The training program determines the communication method (worksheet, test, verbal communication, or other means deemed appropriate) and whether the corrective procedures for these tasks are actually performed.
diagnose	To locate the root cause or nature of a problem by using the specified procedure.
disassemble	To separate a component's parts as a preparation for cleaning, inspection, or service.
find	To locate a particular problem, e.g., shorts, grounds or opens in an electrical circuit.
high voltage	Voltages of 50 volts or higher.
identify	To establish the identity of a vehicle or component prior to service; to determine the nature or degree of a problem.
inspect	(see *check*)
install (reinstall)	To place a component in its proper position in a system.
jump-start	To use an auxiliary power supply, i.e., battery, battery charger, etc. to assist a battery to crank an engine.
leak test	To locate the source of leaks in a component or system.
listen	To use audible clues in the diagnostic process; to hear the customer's description of a problem.
lubricate	To employ the correct procedures and materials in performing the prescribed service.
on-board diagnostics (OBD)	A diagnostic system contained in the Powertrain Control Module (PCM), which monitors computer inputs and outputs for failures. OBD II is an industry-standard, second generation OBD system that monitors emissions control systems for degradation as well as failures.
parasitic draw	Electrical loads which are still present when the circuit is turned OFF.
perform	To accomplish a procedure in accordance with established methods and standards.
perform necessary action	Indicates that the student is to perform the diagnostic routine(s) and perform the corrective action item. Where various scenarios (conditions or situations) are presented in a single task, at least one of the scenarios must be accomplished.
pressure test	To use air or fluid pressure to determine the integrity, condition, or operation of a component or system.
priority ratings	Indicates the minimum percentage of tasks, by area, a program must include in its curriculum in order to be certified in that area.
reassemble	(see *assemble*)
remove	To disconnect and separate a component from a system.
repair	To restore a malfunctioning component or system to operating condition.
replace	To exchange an unserviceable component with a new or rebuilt component; to reinstall a component.
service	To perform a specified procedure when called for in the owner's or service manual.

test	To verify condition through the use of meters, gauges, or instruments.
torque	To tighten a fastener to specified degree of tightness (in a given order or pattern if multiple fasteners are involved on a single component).
vacuum test	To determine the integrity and operation of a vacuum (negative pressure) operated component and/or system.
verify	To establish that a problem exists after hearing the customer's complaint and performing a preliminary diagnosis.
voltage drop	A reduction in voltage (electrical pressure) caused by the resistance in a component or circuit.

ENGINE PERFORMANCE TOOLS AND EQUIPMENT

Many different tools and many kinds of testing and measuring equipment are used to service engine performance systems. NATEF has identified many of these and has said an Engine Performance technician must know what they are and how and when to use them. The tools and equipment listed by NATEF are covered in the following discussion. Also included are the tools and equipment you will use while completing the job sheets. Although you need to be more than familiar with and will be using common hand tools, they are not part of this discussion. You should already know what they are and how to use and care for them.

Scan Tools

The introduction of computer-controlled systems brought with it the need for tools capable of troubleshooting electronic control systems. There is a variety of scan tools, available today, that do just that. A scan tool is a microprocessor designed to communicate with the vehicle's computer and is the first diagnostic tool you should use on computerized engine controls. Connected to the computer through diagnostic connectors, a scan tool can access trouble codes, run tests to check system operations, and monitor the activity of the system. Trouble codes and test results are displayed on an LED screen, or printed out on the scanner printer.

Scan tools will retrieve fault codes from a computer's memory and digitally display these codes on the tool. A scan tool may also perform many other diagnostic functions depending on the year and make of the vehicle. Most aftermarket scan tools have removable modules that are updated each year. These modules are designed to test the computer systems on various makes of vehicles. For example, some scan testers have a 3-in-1 module that tests the computer systems on Chrysler, Ford, and General Motors vehicles. A 10-in-1 module is also available to diagnose computer systems on vehicles imported by 10 different manufacturers. These modules plug into the scan tool.

Scan tools are capable of testing many onboard computer systems, such as climate controls, transmission controls, engine computers, antilock brake computers, air bag computers, and suspension computers, depending on the year and make of the vehicle and the type of scan tester. In many cases, the technician must select the computer system to be tested with the scanner after it has been connected to the vehicle.

The scan tool is connected to specific diagnostic connectors on various vehicles. Most manufacturers have one diagnostic connector. This connects the data wire from each onboard computer to a specific terminal in this connector. Other vehicle manufacturers have several different diagnostic connectors on each vehicle, and each of these connectors may be connected to one or more onboard computers. A set of connectors is supplied with the scanner to allow tester connection to various diagnostic connectors on different vehicles.

The scanner must be programmed for the model year, make of vehicle, and type of engine. With some scan tools, this selection is made by pressing the appropriate buttons on the tester, as directed by the digital tester display. On other scan testers, the appropriate memory card must be installed in the tester for the vehicle being tested. Some scan testers have a built-in printer to print test results, while other scan testers may be connected to an external printer.

As automotive computer systems become more complex, the diagnostic capabilities of scan testers continue to expand. Many scan testers now have the capability to store, or "freeze," data into the tester during a road test, and then play back this data when the vehicle is returned to the shop.

Some scan testers now display diagnostic information based on the fault code in the computer memory. Service bulletins published by the manufacturer of the scan tester may be indexed by the tester after the vehicle information is entered in the tester. Other scan testers display sensor specifications for the vehicle being tested.

The vehicle's computer sets trouble codes when a voltage signal is entirely out of its normal range. The codes help technicians identify the cause of the problem when this is the case. If a signal is within its normal range but it still not correct, the vehicle's computer will not display a trouble code. However, a problem may still exist.

With OBD-II, the diagnostic connectors are located in the same place on all vehicles. Also, any scan tools designed for OBD-II will work on all OBD-II systems, therefore the need to have designated scan tools or cartridges is eliminated. The OBD-II scan tool has the ability to run diagnostic tests on all systems and has "freeze frame" capabilities.

Engine Analyzers

When performing a complete engine performance analysis, an engine analyzer is used (Figure 2). An engine analyzer houses all of the necessary test equipment. Although the term *engine analyzer* is often loosely applied to any multipurpose test meter, a complete engine analyzer will incorporate most, if not all, of the test instruments needed to diagnose driveability problems. Most engine analyzers are based on a computer that guides a technician through the tests. Most have the following diagnostic tools: compression gauge, exhaust analyzer, pressure and vacuum gauges, voltmeter, ohmmeter, vacuum pump, ammeter, tachometer, oscilloscope, timing light and probe, and, of course a scan tool.

With an engine analyzer, you can perform tests on the battery, starting system, charging system, primary and secondary ignition circuits, electronic control systems, the fuel system, the emissions system, and the engine assembly. The analyzer is connected to these systems by a variety of leads, inductive clamps, probes, and connectors. The data received from these connections is processed by several computers within the analyzer.

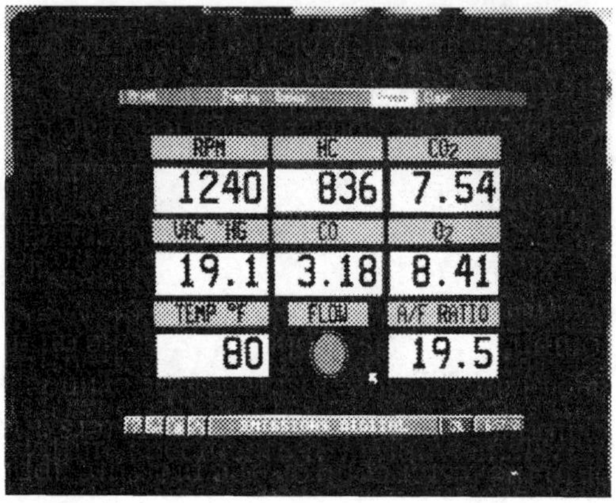

Figure 2 One of many screens on an engine analyzer.

Commands and specifications can be entered into the analyzer on a keyboard. Specifications, commands, and test results are displayed on the CRT screen. Some analyzers will graphically display test results on their CRT screen. The analyzer's printer can also print out copies of the information that appears on the screen.

Most engine analyzers have both manual and automatic test modes. In the manual modes, any single test, such as cylinder compression or alternator output, can be performed. The manual test mode is useful when there is little need to do a complete test. The automatic test mode is useful when looking at general performance. When the automatic test mode is selected, specific tests are automatically performed in a specific sequence.

The analyzer may compare all the test results to the vehicle manufacturer's specifications. When the test series is completed, the analyzer prints a report indicating those readings that were not within specifications. Many analyzers also provide diagnostic assistance for the problems indicated by the readings that were not within specifications.

Some engine analyzers have vehicle specifications on a computer, and the technician enters the necessary information, such as the VIN, model year, and engine size, for the vehicle being tested. Specifications may be updated simply by obtaining a new disk from the equipment manufacturer.

A phone modem is contained in some engine analyzers to provide networking capabilities. This phone modem allows the technician to unload all the technical reports and pattern reports of a specific problem vehicle to off-location technical support teams.

Exhaust Analyzers

Federal laws require that new cars and light trucks must meet specific emissions levels. State governments have also passed laws requiring that car owners maintain their vehicles so that the emissions remain below an acceptable level. Most states require an annual emissions inspection to meet that goal. Many shops have an exhaust analyzer for inspection purposes.

Exhaust analyzers (Figure 3) are also very valuable diagnostic tools. By looking at the quality of an engine's exhaust, a technician is able to observe the effects of the combustion process. Any defect can cause a change in exhaust quality. The amount and type of change serves as the basis of diagnostic work.

Early emission analyzers measured the amount of hydrocarbons (HC) and carbon monoxides (CO). Exhaust analyzers normally measure HC in parts per million (ppm) or grams per mile (g/mi). CO is measured as a percent of the total exhaust.

Many of the emission control devices that have been added to vehicles over the past 30 years have decreased the amount of HC and CO in the exhaust. This is especially true of catalytic converters. These devices alter the contents of the exhaust. Therefore, checking the HC and CO contents in the exhaust may not be a true indication of the operation of an engine.

Figure 3 An exhaust gas analyzer.

The manufacturers of exhaust analyzers have altered their machines so that they can look at the efficiency of an engine, in spite of the effectiveness of the emission controls. These machines are four-gas exhaust analyzers. In addition to measuring HC and CO levels, a four-gas exhaust analyzer also monitors carbon dioxide (CO_2) and oxygen (O_2) levels in the exhaust. The latter two gases are changed only slightly by the emission controls and therefore can be used to check engine efficiency. Many exhaust analyzers are also available that measure a fifth gas, oxides of nitrogen (NO_x).

By measuring oxides of nitrogen (NO_x), carbon dioxide (CO_2), and oxygen (O_2), in addition to HC and CO, a technician gets a better look at the efficiency of the engine. Keep in mind that an exhaust analyzer is an excellent diagnostic tool and is not used just for comparing emissions levels against standards. There is a desired relationship between the five gases. Any deviation from this relationship can be used to diagnose a driveability problem.

Ideally, the combustion process will combine fuel (HC) and oxygen (O_2) to form water and carbon dioxide (CO_2). Although most of the air brought into the engine is nitrogen, this gas should not become part of the combustion process and should pass out of the exhaust as nitrogen. NO_x is formed only when there are very high combustion temperatures. High amounts of NO_x can be caused by cooling system problems, carbon build-up on the pistons (causing higher than normal compression ratios), a defective EGR system, and lean mixtures.

Small amounts of CO_2 are normally present in our air and CO_2 is not considered a pollutant. CO_2 is a desired element in the exhaust stream and is only present when there is complete combustion. Therefore, the more CO_2 in the exhaust stream, the better.

O_2 is used to oxidize CO and HC into water and CO_2. The O_2 level is an indicator of air/fuel mixture. As the mixture goes lean, O_2 levels increase. As the mixture goes rich, O_2 moves and stays low. Ideally, we would like to see very low amounts of O_2 in the exhaust stream.

Vacuum Gauge

Measuring intake manifold vacuum is another way to diagnose the condition of an engine. Manifold vacuum is tested with a vacuum gauge. Vacuum is formed on a piston's intake stroke. As the piston moves down, it lowers the pressure of the air in the cylinder—if the cylinder is sealed. This lower cylinder pressure is called engine vacuum. If there is a leak, atmospheric pressure will force air into the cylinder and the resultant pressure will not be as low. The reason atmospheric pressure enters is simply that whenever there is a low and high pressure, the high pressure will always move toward the low pressure. Vacuum is measured in inches of mercury (in./Hg) and in kilopascals (kPa) or millimeters of mercury (mm/Hg).

To measure vacuum, a flexible hose on the vacuum gauge (Figure 4) is connected to a source of manifold vacuum, either on the manifold or a point below the throttle plates. The test is made with the engine cranking or running. A good vacuum reading is typically at least 16 in./Hg. However, a reading of 15 to 20 in./Hg (50 to 65 kPa) is normally acceptable. Since the intake stroke of each cylinder occurs at a different time, the production of vacuum occurs in pulses. If the amount of vacuum produced by each cylinder is the same, the vacuum gauge will show a steady reading. If one or more cylinders are producing different amounts of vacuum, the gauge will show a fluctuating reading.

Figure 4 A vacuum gauge.

Circuit Tester

Circuit testers are used to identify short and open circuits in any electrical circuit. Low-voltage testers are used to troubleshoot 6- to 12-volt circuits. A circuit tester, commonly called a test light, looks like a stubby ice pick. Its handle is transparent and contains a light bulb. A probe extends from one end of the handle and a ground clip and wire from the other end. When the ground clip is attached to a good ground and the probe touched to a live connector, the bulb in the handle will light up. If the bulb does not light, voltage is not available at the connector.

WARNING: *Do not use a conventional 12-V test light to diagnose components and wires in electronic systems. The current draw of these test lights may damage computers and system components. High-impedance test lights are available for diagnosing electronic systems.*

A self-powered test light is called a continuity tester. It is used on non-powered circuits. It looks like a regular test light, except that it has a small internal battery. When the ground clip is attached to the negative side of a component and the probe touched to the positive side, the lamp will light if there is continuity in the circuit. If an open circuit exists, the lamp will not light. Do not use any type of test light or circuit tester to diagnose automotive air bag systems.

Voltmeter

A voltmeter has two leads: a red positive lead and a black negative lead. The red lead should be connected to the positive side of the circuit or component. The black should be connected to ground or to the negative side of the component. Voltmeters should be connected across the circuit being tested.

The voltmeter measures the voltage available at any point in an electrical system. A voltmeter can also be used to test voltage drop across an electrical circuit, component, switch, or connector. A voltmeter can also be used to check for proper circuit grounding.

Ohmmeter

An ohmmeter measures the resistance to current flow in a circuit. In contrast to the voltmeter, which uses the voltage available in the circuit, the ohmmeter is battery powered. The circuit being tested must be open. If the power is on in the circuit, the ohmmeter will be damaged.

The two leads of the ohmmeter are placed across or in parallel with the circuit or component being tested. The red lead is placed on the positive side of the circuit and the black lead is placed on the negative side of the circuit. The meter sends current through the component and determines the amount of resistance based on the voltage dropped across the load. The scale of an ohmmeter reads from zero to infinity. A zero reading means there is no resistance in the circuit and may indicate a short in a component that should show a specific resistance. An infinite reading indicates a number higher than the meter can measure. This usually is an indication of an open circuit.

Ohmmeters are also used to trace and check wires or cables. Assume that one wire of a four-wire cable is to be found. Connect one probe of the ohmmeter to the known wire at one end of the cable and touch the other probe to each wire at the other end of the cable. Any evidence of resistance, such as meter needle deflection, indicates the correct wire. Using this same method, you can check a suspected defective wire. If resistance is shown on the meter, the wire is sound. If no resistance is measured, the wire is defective (open). If the wire is okay, continue checking by connecting the probe to other leads. Any indication of resistance indicates that the wire is shorted to one of the other wires and that the harness is defective.

Ammeter

An ammeter measures current flow in a circuit. The ammeter must be placed into the circuit or in series with the circuit being tested. Normally, this requires disconnecting a wire or connector from a component and connecting the ammeter between the wire or connector and the component. The red lead of the ammeter should always be connected to the side of the connector closest to the positive side of the battery and the black lead should be connected to the other side.

It is much easier to test current using an ammeter with an inductive pickup. The pickup clamps around the wire or cable being tested. These ammeters measure amperage based on the magnetic field created by the current flowing through the wire. This type of pickup eliminates the need to separate the circuit to insert the meter.

Because ammeters are built with very low internal resistance, connecting them in series does not add any appreciable resistance to the circuit. Therefore, an accurate measurement of the current flow can be taken.

Volt/Ampere Tester

A volt/ampere tester (VAT) is used to test batteries, starting systems, and charging systems. The tester contains a voltmeter, ammeter, and carbon pile. The carbon pile is a variable resistor. A knob on the tester allows the technician to vary the resistance of the pile. When the tester is attached to the battery, the carbon pile will draw current out of the battery. The ammeter will read the amount of current draw. When testing a battery, the resistance of the carbon pile is adjusted so the current draw matches the ratings of the battery.

Logic Probes

In some circuits pulsed or digital signals pass through the wires. These "on-off" digital signals either carry information or provide power to drive a component. Many sensors used in a computer-control circuit send digital information back to the computer. To check the continuity of the wires that carry digital signals, a logic probe can be used.

A logic probe has three different colored LEDs. A red LED lights when there is high voltage at the point being probed. A green LED lights to indicate low voltage. And a yellow LED indicates the presence of a voltage pulse. The logic probe is powered by the circuit and reflects only the activity at the point being probed. When the probe's test leads are attached to a circuit, the LEDs display the activity.

If a digital signal is present, the yellow LED will turn on. When there is no signal, the LED is off. If voltage is present, the red or green LEDs will light, depending on the amount of voltage. When there is a digital signal and the voltage cycles from low to high, the yellow LED will be lit and the red and green LEDs will cycle indicating a change in the voltage.

DMMs

It's not necessary for a technician to own separate meters to measure volts, ohms, and amps, because a multimeter can be used instead. Top-of-the-line multimeters are multifunctional. Most test volts, ohms, and amperes in both DC and AC. Usually there are several test ranges provided for each of these functions. In addition to these basic electrical tests, multimeters also test engine rpm, duty cycle, pulse width, diode condition, frequency, and even temperature. The technician selects the desired test range by turning a control knob on the front of the meter.

Multimeters are available with either analog or digital displays. But, the most commonly used multimeter is the digital volt/ohmmeter (DVOM), which is often referred to as a digital multimeter (DMM). There are several drawbacks to using analog-type meters for testing electronic control systems. Many electronic components require very precise test results. Digital meters can measure volts, ohms, or amperes in tenths and hundredths. Another problem with analog meters is their low internal resistance (input impedance). The low input impedance allows too much current to flow through circuits and should not be used on delicate electronic devices.

Digital meters, on the other hand, have high input impedance, usually at least 10 megohms (10 million ohms). Metered voltage for resistance tests is well below 5 volts, reducing the risk of damage to sensitive components and delicate computer circuits. A high-impedance digital multimeter must be used to test the voltage of some components and systems such as an oxygen (O_2) sensor circuit. If a low-impedance analog meter is used in this type of circuit, the current flow through the meter is high enough to damage the sensor.

DMMs either have an "auto range" feature, in which the meter automatically selects the appropriate scale, or they must be set to a particular range. In either case, you should be familiar with the ranges and the different settings available on the meter you are using. To designate particular ranges and readings, meters display a prefix before the reading or range. If the meter has a setting for mAmps, this means the readings will be given in milli-amps or 1/1000th of an amp. Ohmmeter scales are expressed as a multiple of tens or use the prefix K or M. K stands for Kilo or 1000. A reading of 10K ohms equals 10,000 ohms. An M stands for Mega or 1,000,000. A reading of 10M ohms equals 10,000,000 ohms. When using a meter with an auto range, make sure you note the range being used by the meter. There is a big difference between 10 ohms and 10,000,000 ohms.

WARNING: *Many digital multimeters with auto range display the measurement with a decimal point.*

After the test range has been selected, the meter is connected to the circuit in the same way as if it were an individual meter.

When using the ohmmeter function, the DMM will show a zero or close to zero when there is good continuity. If the continuity is very poor, the meter will display an infinite reading. This reading is usually shown as a blinking "1.000", a blinking "1", or an "OL". Before taking any measurement, calibrate the meter. This is done by holding the two leads together and adjusting the meter reading to zero. Not all meters need to be calibrated; some digital meters automatically calibrate when a scale is selected. On meters that require calibration, it is recommended that the meter be zeroed after changing scales.

Multimeters may also have the ability to measure duty cycle, pulse width, and frequency. All of these represent voltage pulses caused by the turning on and off of a circuit or the increase and decrease of voltage in a circuit. Duty cycle is a measurement of the amount of time something is on compared to the time of one cycle and is measured in a percentage.

Pulse width is similar to duty cycle except that it is the exact time something is turned on and is measured in milliseconds. When measuring duty cycle, you are looking at the amount of time something is on during one cycle.

The number of cycles that occur in one second is called the frequency. The higher the frequency, the more cycles occur in a second. Frequencies are measured in Hertz. One Hertz is equal to one cycle per second.

Lab Scopes

An oscilloscope is a visual voltmeter. An oscilloscope converts electrical signals to a visual image representing voltage changes over a specific period of time. This information is displayed in the form of a continuous voltage line called a waveform pattern or trace.

An oscilloscope screen is a cathode ray tube (CRT), which is very similar to the picture tube in a television set. High voltage from an internal source is supplied to an electron gun in the back of the CRT when the oscilloscope is turned on. This electron gun emits a continual beam of electrons against the front of the CRT. The external leads on the oscilloscope are connected to deflection plates above and below and on each side of the electron beam. When a voltage signal is supplied from the external leads to the deflection plates, the electron beam is distorted and strikes the front of the screen in a different location to indicate the voltage signal from the external leads.

An upward movement of the voltage trace on an oscilloscope screen indicates an increase in voltage, and a downward movement of this trace represents a decrease in voltage. As the voltage trace moves across an oscilloscope screen, it represents a specific length of time.

The size and clarity of the displayed waveform is dependent on the voltage scale and the time reference selected. Most scopes are equipped with controls that allow voltage and time interval selection. It is important, when choosing the scales, to remember that a scope displays voltage over time.

Dual-trace oscilloscopes can display two different waveform patterns at the same time.

With a scope, precise measurement is possible. A scope will display any change in voltage as it occurs. This is especially important for diagnosing intermittent problems. It is also invaluable for checking the primary and secondary ignition circuits.

The screen of a lab scope is divided into small divisions of time and voltage. Time is represented by the horizontal movement of the waveform. Voltage is measured with the vertical position of the waveform. Since the scope displays voltage over time, the waveform moves from the left (the beginning of measured time) to the right (the end of measured time). The value of the divisions can be adjusted to improve the view of the voltage waveform.

Since a scope displays actual voltage, it will display any electrical noise or disturbances that accompany the voltage signal. Noise is primarily caused by radio frequency interference (RFI), which may come from the ignition system. RFI is an unwanted voltage signal that rides on a signal. This noise can cause intermittent problems with unpredictable results. The noise causes slight increases and decreases in the voltage. When a computer receives a voltage signal with noise, it will try to react to the minute changes. As a result, the computer responds to the noise rather than the voltage signal.

Static Safeguards

Some manufacturers mark certain components and circuits with a code or symbol to warn technicians that they are sensitive to electrostatic discharge. Static electricity can destroy or render a component useless.

When handling any electronic part, especially those that are static sensitive, follow the guidelines below to reduce the possibility of electrostatic build-up on your body and the inadvertent discharge to the electronic part. If you are not sure if a part is sensitive to static, treat it as if it were.

1. Always touch a known good ground before handling the part. This should be repeated while handling the part and more frequently after sliding across a seat, sitting down from a standing position, or walking a distance.
2. Avoid touching the electrical terminals of the part, unless you are instructed to do so in the written service procedures. It is good practice to keep your fingers off all electrical terminals, as the oil from your skin can cause corrosion.
3. When you are using a voltmeter, always connect the negative meter lead first.
4. Do not remove a part from its protective package until it is time to install the part.
5. Before removing the part from its package, ground yourself and the package to a known good ground on the vehicle.

Some tool manufacturers have grounding straps that are available. These are designed to fit on you, at one end, and be fastened to a good ground on the other end. They are some of the best safeguards against static electricity.

Battery Hydrometer

On unsealed batteries, the specific gravity of the electrolyte can be measured to give a fairly good indication of the battery's state of charge. A hydrometer is used to perform this test. A battery hydrometer consists of a glass tube or barrel, rubber bulb, rubber tube, and a glass float or hydrometer with a scale built into its upper stem. The glass tube encases the float and forms a reservoir for the test electrolyte. Squeezing the bulb pulls electrolyte into the reservoir.

When filled with test electrolyte, the sealed hydrometer float bobs in the electrolyte. The depth to which the glass float sinks in the test electrolyte indicates its relative weight compared to water. The reading is taken off the scale by sighting along the level of the electrolyte.

If the hydrometer floats deep in the electrolyte, the specific gravity is low. If the hydrometer floats shallow in the electrolyte, the specific gravity is high.

At extremely high and low electrolyte temperatures, it is necessary to correct the reading by adding or subtracting 4 points (0.004) for each 10 degrees Fahrenheit above or below the standard of 80°F. Most hydrometers have a built-in thermometer to measure the temperature of the electrolyte. The hydrometer reading can be misleading if it is not adjusted. It is important to make these adjustments at high and low temperatures to determine the battery's true state of charge.

On many sealed maintenance-free batteries a special temperature-compensated hydrometer is built into the battery cover. A quick visual check indicates the battery's state of charge. The hydrometer has a green ball within a cage that is attached to a clear plastic rod. The green ball floats at a predetermined specific gravity of the electrolyte that represents about a 65% state of charge. When the green ball floats, it rises within the cage and positions itself under the rod. Visually, a green dot then shows in the center of the hydrometer. In testing, the green dot means the battery is charged enough for testing. If the green dot is not visible and has a dark appearance, it means the battery must be charged before the test procedure is performed.

While charging, the appearance of the green dot means that the battery is sufficiently charged. Charging can be stopped to prevent overcharging.

It is important when observing the hydrometer that the battery has a clean top so that you can see the correct indication. A flashlight may be required in some poorly lit areas. Always look straight down when viewing the hydrometer.

On some special applications, some hydrometers feature a red dot indication in addition to the green dot, dark, and clear appearances. The red dot means the battery is nearing complete discharge and must be charged before being used.

Tachometer

A tachometer is used to measure engine speed. Like other meters, tachometers are available in analog and digital types. Digital meters are the most common. Tachometers are connected to the ignition system to monitor ignition pulses. These pulses are then converted to engine speed by the meter.

Several types of inductive pickup tachometers that simplify rpm testing are now available. An inductive tachometer simply clamps over the number one spark plug wire. The digital display gives the engine rpm, based on the magnetic pulses created by the secondary voltage in the wire. This type of tachometer is suitable for distributorless ignition systems.

Another type of digital tachometer is the photoelectric tachometer. The photoelectric tachometer converts light pulses into rpm readings. This type of meter has an internal light source, powered either by an internal battery or by the car's battery. Reflective tape is applied to any rotating part of the engine, such as the crankshaft pulley. The photoelectric cell in the tachometer senses the reflected light each time the reflective tape rotates through the meter's light beam. An engine rpm calculation is made by the tachometer from these reflected light pulses.

Magnetic pulse sensors are used to determine crankshaft position and rotational speed. The sensors were usually mounted over the harmonic balancer, the flywheel, beside a crankshaft-mounted reluctor, or inside the distributor. As the mechanical components rotate, voltage pulses are magnetically induced in the sensor. An electronic control module monitors the voltage pulses to determine crankshaft position and engine speed.

Tach-Dwellmeter

A tool that was commonly used on older engines was a tach-dwellmeter. This meter is a combination tachometer and dwellmeter. A tachometer measures engine speed. Engine speed is measured in rpms (revolutions per minute). A dwellmeter measures the time a circuit is on. On ignition systems, the dwell time is the degree of crankshaft rotation during which the primary circuit is on. A small internal dry-cell battery powers most tachdwellmeters. The red tach-dwellmeter lead is connected to the negative primary coil terminal, and the black meter lead is connected to ground. A switch on the meter must be set in the rpm, or dwell, position. On distributorless ignition systems a special tachometer lead may be provided in the ignition system for the tach-dwellmeter connection.

Timing Light

A timing light checks ignition timing in relation to crankshaft position. Two leads on the timing light must be connected to the battery terminals with the correct polarity. Most timing lights have an inductive clamp that fits over the number one spark plug wire. Older timing lights have a lead that goes in series between the number one spark plug wire and the spark plug. A trigger on the timing light acts as an off/on switch. When the trigger is pulled with the engine running, the timing light emits a beam of light each time the spark plug fires.

To check the ignition timing, the flashing timing light is aimed at the ignition timing marks. The timing marks are usually located on the crankshaft pulley or on the flywheel. A stationary pointer, line, or notch is positioned above the rotating timing marks. The timing marks are lines on the crankshaft pulley or flywheel that represent various degrees of crankshaft rotation when the number one piston is before top dead center (BTDC) on the compression stroke. The TDC crankshaft position and the degrees are usually identified in the group of timing marks. Some timing marks include degree lines representing the after top dead center (ATDC) crankshaft position.

When the number one piston is at TDC on the compression stroke, the line or notch will line up with the zero reference mark on the timing plate. Usually an engine is timed so the number one spark plug fires several degrees BTDC.

When the timing light is connected to the battery and the number one spark plug wire, it will flash every time the number one spark plug fires. When pointed at the timing marks, the strobe effect of the light will freeze the spinning timing marks as they pass the timing scale. The ignition timing is checked by observing the degree of crankshaft rotation BTDC or ATDC when the spark plug fires.

Before checking the ignition timing, complete all of the vehicle manufacturer's recommended procedures. On fuel injected engines, special timing procedures are required, such as disconnecting a timing connector to be sure the computer does not provide any spark advance while checking basic ignition timing. These timing instructions are usually provided on the underhood emission label. Most late-model engines have preset ignition timing and adjustments cannot be made.

Base timing is checked at base, or curb, idle. Advanced timing is checked at a higher rpm specified by the vehicle's manufacturer. As engine speed increases, the timing is advanced (mechanically, by vacuum, or electronically). The actual amount of timing advance can be measured with an advance timing light.

Many timing lights have a timing advance knob that may be used to check spark advance. This knob has an index line and a degree scale surrounding the knob. Prior to checking the spark advance, the basic timing should be checked. Then accelerate the engine to 2,500 rpm or the speed recommended by the vehicle manufacturer. While maintaining this rpm, slowly rotate the advance knob toward the advanced position until the timing marks come back to the basic timing position. Under this condition, the index mark on the advance knob is pointing to the number of degrees advance provided by the computer or distributor advances. The reading on the degree scale may be compared to the vehicle manufacturer's specifications to determine if the spark advance is correct.

A more versatile advanced timing light is the digital timing light. This type of timing light electronically measures timing advance as the engine rpm is increased and displays timing advance on an LED display. This light flashes only when a trigger is squeezed. When the trigger is released, the LED displays engine rpm. This combined feature eliminates the need for a separate tachometer when setting timing.

Magnetic Timing Probe

A timing light is not necessary for tuning many cars. A magnetic probe receptacle on the crankshaft position sensor allows the ignition timing to be electronically monitored with a magnetic timing probe. Changes in the magnetic field create electrical pulses in the tip of the timing probe. These pulses are monitored by the timing meter to determine crankshaft position and ignition timing. The timing indicated by this type of equipment is extremely accurate.

The only trick to using a magnetic timing probe is correcting the offset of the probe receptacle. The receptacle is usually situated on either side of the vehicle's coil pickup. The degree of offset must be factored into the timing probe's reading or the timing readout will be inaccurate by that many degrees. The timing meter must be programmed for the degree of offset specified by the car manufacturer.

Electronic Ignition Module Tester

A defective electronic ignition module can cause a variety of problems. Some may be as obvious as a no-spark, no-start condition. Others may be only intermittent problems, such as a misfire at cruising speeds or under certain load or temperature conditions. Tracing the source of intermittent problems to the ignition module can be difficult without an ignition module tester.

An electronic module ignition tester evaluates and determines if the module is operating within a given set of design parameters. It does so by simulating normal operating conditions while looking for faults in key components.

Feeler Gauge

A feeler gauge is a thin strip of metal or plastic of known and closely controlled thickness. Several of these strips are often assembled together as a feeler gauge set that looks like a pocketknife. The desired thickness gauge can be pivoted away from the others for convenient use. A feeler gauge set usually contains strips or leaves of 0.002- to 0.010-inch thickness (in steps of 0.001 inch) and leaves of 0.012- to 0.024-inch thickness (in steps of 0.002 inch).

A feeler gauge can be used by itself to measure clearances and end play. Round, wire feeler gauges are often used to measure spark plug gap. The round gauges are designed to give a better feel for the fit of the gauge in the gap.

A specially designed tool measures and corrects the gap. This pliers-like tool utilizes flat gauges for the adjustment procedure. The gauges are mounted on the tool like spokes on a wheel. Above the gauges is an anvil, which is used to apply pressure to the electrode. On the opposite end of the tool is a curved seat. This seat performs two functions: it supports the plug shell during the procedure and also compresses the ground electrode against the gauge, thus setting the air gap.

A tapered regapping tool is simply a piece of tapered steel with leading and trailing edges of different dimensions. Between these two points, the gauge varies in thickness. A scale, located above the gauge, indicates the thickness at any given point. When the gauge is slid between electrodes, it stops when the air gap size reaches the thickness on the gauge. The scale reading is made in thousandths of an inch. Adjusting slots are available to bend the ground electrode as needed to make the air gap adjustment.

Torque-Indicating Wrench

Torque is the twisting force used to turn a fastener against the friction between the threads and between the head of the fastener and the surface of the component. The fact that practically every vehicle and engine manufacturer publishes a list of torque recommendations is ample proof of the importance of using proper amounts of torque when tightening nuts or bolts. The amount of torque applied to a fastener is measured with a torque-indicating or torque wrench.

There are three basic types of torque-indicating wrenches available with pounds per inch and pounds per foot increments: a beam torque wrench that has a beam that points to the torque reading, a "click"-type torque wrench in which the desired torque reading is set on the handle (when the torque reaches that level, the wrench clicks), and a dial torque wrench that has a dial that indicates the torque exerted on the wrench. Some designs of this type torque wrench have a light or buzzer that turns on when the desired torque is reached.

An important item that many forget to torque properly is the spark plug. A spark plug is a plug that allows a spark to enter the combustion chamber without allowing combustion gases to leak. The latter purpose is the reason spark plugs need to be properly tightened in their bores.

Spark Plug Thread Repair

Sometimes when spark plugs are removed from a cylinder head, the threads have traces of metal on them. This happens more often with aluminum heads. When this occurs, the spark plug bore must be corrected by installing thread inserts.

When installing spark plugs, if the plugs cannot be installed easily by hand, the threads in the cylinder head may need to be cleaned with a thread-chasing tap. Be especially careful not to cross-thread the plugs when working with aluminum heads. Always tighten the plugs with a torque wrench and the correct spark plug socket, following the vehicle manufacturer's specifications.

Spark Tester

A spark tester is a fake spark plug. The tester is constructed like a spark plug but doesn't have a ground electrode. In place of the electrode there is a grounding clamp. Using test spark plugs is an easy way to determine if the ignition problem is caused by something in the primary or secondary circuit.

The spark tester is inserted in the spark plug end of an ignition cable. When the engine is cranked, a spark should be seen from the tester to a ground. Experience with these testers will also help you determine the intensity of the spark.

Sensor Tools

Oxygen sensors are replaced as part of the preventative maintenance program and when they are faulty. Because they are shaped much like a spark plug with wires or a connector coming out of the top, ordinary sockets don't fit well. For this reason, tool manufacturers provide special sockets for these sensors.

Special sockets are also available for other sending units and sensors.

Fuel Pressure Gauge

A fuel pressure gauge (Figure 5) is very important for diagnosing fuel injection systems. These systems rely on very high fuel pressures, from 35 to 70 psi. A drop in fuel pressure will reduce the amount of fuel delivered to the injectors and result in a lean air/fuel mixture.

A fuel pressure gauge is used to check the discharge pressure of fuel pumps, the regulated pressure of fuel injection systems, and injector pressure drop. This test can identify faulty pumps, regulators, or injectors and can identify restrictions present in the fuel delivery system. Restrictions are typically caused by a dirty fuel filter, collapsed hoses, or damaged fuel lines.

Figure 5 A fuel pressure gauge.

Some fuel pressure gauges also have a valve and outlet hose for testing fuel pump discharge volume. The manufacturer's specification for discharge volume will be given as a number of pints or liters of fuel that should be delivered in a certain number of seconds.

Injector Balance Tester

The injector balance tester (Figure 6) is used to test the injectors in a port fuel injected engine for proper operation. A fuel pressure gauge is also used during the injector balance test. The injector balance tester contains a timing circuit, and some injector balance testers have an off-on switch. A pair of leads on the tester must be connected to the battery with the correct polarity. The injector terminals are disconnected, and a second double lead on the tester is attached to the injector terminals.

Before conducting an injector test, a fuel pressure gauge is connected to the Schrader valve on the fuel rail, and the ignition switch should be cycled two or three times until the specified fuel pressure is indicated on the pressure gauge. When the tester's push button is depressed, the tester energizes the injector winding for a specific length of time, and the technician records the pressure decrease on the fuel pressure gauge. This procedure is repeated on each injector.

Figure 6 A fuel injector balance tester.

Some vehicle manufacturers provide a specification of 3-psi (20 kPa) maximum difference between the pressure readings after each injector is energized. If the injector orifice is restricted, there is not much pressure decrease when the injector is energized. Acceleration stumbles, engine stalling, and erratic idle operation are caused by restricted injector orifices. The injector plunger is sticking open if excessive pressure drop occurs when the injector is energized. Sticking injector plungers may result in a rich air/fuel mixture.

Because electronic fuel injection systems are pressurized, make sure you depressurize the system before opening up the system for diagnostics or repair work.

Injector Circuit Test Light

A special test light called a "noid light" can be used to determine if a fuel injector is receiving its proper voltage pulse from the computer. The wiring harness connector is disconnected from the injector and the noid light is plugged into the connector. After disabling the ignition to prevent starting, the engine is turned over by the starter motor. The noid light will flash rapidly if the voltage signal is present. No flash usually indicates an open in the power feed or ground circuit to the injector.

Fuel Injector Cleaners

Fuel injectors spray a certain amount of fuel into the intake system. If the fuel pressure is low, not enough fuel will be sprayed. This is also true if the fuel injector is dirty. Normally clogged injectors are the result of inconsistencies in gasoline detergent levels and the high sulfur content of gasoline. When these sensitive fuel injectors become partially clogged, fuel flow is restricted. Spray patterns are altered, causing poor performance and reduced fuel economy.

The solution to a sulfated and/or plugged fuel injector is to clean it, not replace it. There are two kinds of fuel injector cleaners (Figure 7). One is a pressure tank. A mixture of solvent and unleaded gasoline is placed in the tank, following the manufacturer's instructions for mixing, quantity, and safe handling. The vehicle's fuel pump must be disabled and, on some vehicles, the fuel line must be blocked between the pressure regulator and the return line. Then, the hose on the pressure tank is connected to the service port in the fuel system. The in-line valve is then partially opened and the engine is started. It should run at approx-

Figure 7 A typical fuel injector cleaning machine.

imately 2,000 rpm for about 10 minutes to clean the injectors thoroughly.

An alternative to the pressure tank is a pressurized canister in which the solvent solution is premixed. Use of the canister-type cleaner is similar to this procedure, but does not require mixing or pumping.

The canister is connected to the injection system's servicing fitting, and the valve on the canister is opened. The engine is started and allowed to run until it dies. Then the canister is discarded.

Fuel Line Tools

Many vehicles are equipped with quick-connect line couplers (Figure 8). These work well to seal the connection but are almost impossible to disconnect if the correct tools are not used. There is a variety of quick-connect fittings and tools.

Vacuum Pump

There are many vacuum-operated devices and switches on cars. These devices use engine vacuum to cause a mechanical action or to switch something on or off. The tool used to test vacuum-actuated components is the vacuum pump. There are two types of vacuum pumps: an electrical operated pump and a hand-held pump (Figure 9). The hand-held pump is most often used for diagnostics. A hand-held vacuum pump consists of a hand pump, a vacuum gauge, and a length of

Figure 8 A quick-connect coupling tool.

Figure 9 A hand-held vacuum pump.

rubber hose used to attach the pump to the component being tested. Tests with the vacuum pump can usually be performed without removing the component from the car.

When the handles of the pump are squeezed together, a piston inside the pump body draws air out of the component being tested. The partial vacuum created by the pump is registered on the pump's vacuum gauge. While forming a vacuum in a component, watch the action of the component. The vacuum level needed to actuate a given component should be compared to the specifications given in the factory service manual.

The vacuum pump is also commonly used to locate vacuum leaks. This is done by connecting the vacuum pump to a suspect vacuum hose or component and applying vacuum. If the needle on the vacuum gauge begins to drop after the vacuum is applied, a leak exists somewhere in the system.

Vacuum Leak Detector

A vacuum or compression leak might be revealed by a compression check, a cylinder leak down test, or a manifold vacuum test. However, finding the location of the leak can often be very difficult.

A simple, but time-consuming way to find leaks in a vacuum system is to check each component and vacuum hose with a vacuum pump. Simply apply vacuum to the suspected area and watch the gauge for any loss of vacuum. A good vacuum component will hold the vacuum that is applied to it.

Another method of leak detection is done by using an ultrasonic leak detector. Air rushing through a vacuum leak creates a high-frequency sound, higher than the range of human hearing. An ultrasonic leak detector is designed to hear the frequencies of the leak. When the tool is passed over a leak, the detector responds to the high-frequency sound by emitting a warning beep. Some detectors also have a series of LEDs that light up as the frequencies are received. The closer the detector is moved to the leak, the more LEDs light up or the faster the beeping occurs. This allows the technician to zero in on the leak. An ultrasonic leak detector can sense leaks as small as 1/500 inch and accurately locate the leak to within 1/16 inch.

An ultrasonic leak detector can also be used to detect the source of compression leaks, bearing wear, and electrical arcing. It can also be used to diagnose fuel injector operation.

Pinch-Off Pliers

The need to pinch off a rubber hose is common during diagnostics and service. Special pliers are designed to do this without damaging the hose. These pliers are much like vise-grip pliers in that they hold their position until they are released. The jaws of the pliers are flat and close in a parallel motion. Both of these features prevent damage to the hose.

Pyrometers

The converter should be checked for its ability to convert CO and HC into CO_2 and water. One of the ways to do this is to do a delta temperature test. To conduct this test, use a hand-held digital pyrometer. By touching the pyrometer probe to the exhaust pipe just ahead of and just behind the converter, there should be an increase of at least 100°F or 8% above the inlet temperature reading as the exhaust gases pass through the converter. If the outlet temperature is the same or lower, nothing is happening inside the converter.

Compression Testers

Engines depend on the compression of the air/fuel mixture to have power output. The compression stroke of the piston compresses the air/fuel mixture within the combustion chamber. If the combustion chamber leaks, some of the air/fuel mixture will escape while it is being compressed, resulting in a loss of power and a waste of fuel. The leaks can be caused by burned valves, a blown head gasket, worn rings, a slipped timing belt or chain, worn valve seats, a cracked head, and other factors.

An engine with poor compression (lower compression pressure due to leaks in the cylinder) will not run correctly. If a symptom suggests that the cause of a problem may be poor compression, a compression test is performed.

A compression gauge is used to check cylinder compression. The dial face on the typical compression gauge indicates pressure in both pounds per square inch (psi) and metric kilopascals (kPa). The range is usually 0 to 300 psi and 0 to 2,100 kPa. There are two basic types of compression gauges: a push-in gauge and a screw-in gauge.

The push-in type has a short stem that is either straight or bent at a 45-degree angle. The stem ends in a tapered rubber tip that fits any size spark plug hole. The rubber tip is placed in the spark plug hole, after the spark plugs have been removed, and held there while the engine is cranked through several compression cycles. Although simple to use, the push-in gauge may give inaccurate readings if it is not held tightly in the hole.

The screw-in gauge has a long, flexible hose that ends in a threaded adapter. This type of compression tester is often used because its flexible hose can reach into areas that are difficult to reach with a push-in type tester. The threaded adapters are changeable and come in several thread sizes to fit 10-mm, 12-mm, 14-mm, and 18-mm diameter holes. The adapters screw into the spark plug holes in place of the spark plugs.

Most compression gauges have a vent valve that allows the gauge's meter to hold the highest pressure reading. Opening the valve releases the pressure when the test is complete.

Cylinder Leakage Tester

If a compression test shows that any of the cylinders are leaking, a cylinder leakage test can be performed to measure the percentage of compression lost and help locate the source of leakage.

A cylinder leakage tester (Figure 10) applies compressed air to a cylinder through the spark plug hole. Before the air is applied to the cylinder, the

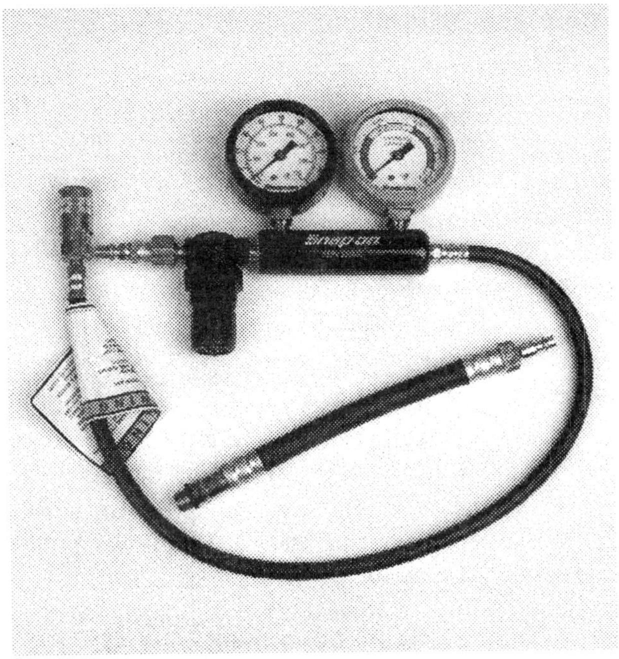

Figure 10 A cylinder leakage tester.

piston of that cylinder must be at TDC on its compression stroke. A threaded adapter on the end of the air pressure hose screws into the spark plug hole. The source of the compressed air is normally the shop's compressed air system. A pressure regulator in the tester controls the pressure applied to the cylinder. An analog gauge registers the percentage of air pressure lost from the cylinder when the compressed air is applied. The scale on the dial face reads 0 to 100 percent.

A zero reading means there is no leakage from the cylinder. Readings of 100 percent would indicate that the cylinder does not hold pressure. The location of the compression leak can be found by listening and feeling around various parts of the engine. If air is felt or heard leaving the throttle plate assembly, a leaking intake valve is indicated. If a bad exhaust valve is responsible for the leakage, air can be felt leaving the exhaust system during the test. Air leaving the radiator would indicate a faulty head gasket or a cracked block or head. If the piston rings are bad, air will be heard leaving the valve cover's breather cap or the oil dipstick tube.

Most vehicles, even new cars, experience some leakage around the rings. Up to 20 percent is considered acceptable during the leakage test. When the engine is actually running, the rings will seal much more tightly and the actual percent of leakage will be lower. However, there should be no leakage around the valves or the head gasket.

Cylinder Power Balance Test

The cylinder power balance test is used to see if all of the engine's cylinders are producing the same amount of power. Ideally, all cylinders will produce the same amount. To check an engine's power balance, the spark plugs for individual cylinders are shorted out one at a time and the change in engine speed is recorded. If all of the cylinders are producing the same amount of power, engine speed will drop the same amount as each plug is shorted. Unequal cylinder power balance can mean a problem in the cylinders themselves, as well as the rings, valves, intake manifold, head gasket, fuel system, or ignition system.

A power balance test is performed quickly and easily using an engine analyzer, because the firing of the spark plugs can be automatically controlled or manually controlled by pushing a button. Some vehicles have a power balance test built into the engine control computer. This test is either part of a routine self-diagnostic operating mode or must be activated by the technician.

On some computer-controlled engines, certain components must be disconnected before attempting the power balance test. Because of the wide variations from manufacturer to manufacturer, always check the appropriate service manual.

Be careful not to run the engine with a shorted cylinder for more than 15 seconds. The unburned fuel in the exhaust can build up in the catalytic converter and create an unsafe situation. Also run the engine for at least 10 seconds between cylinder shortings.

Oil Pressure Gauge

Checking the engine's oil pressure will give you information about engine wear and the condition of the oil pump, pressure regulator, and the entire lubrication system. Lower than normal oil pressures can be caused by excessive engine bearing clearances. Oil pressure is checked at the sending unit passage with an externally mounted mechanical oil pressure gauge. Various fittings are usually supplied with the oil pressure gauge to fit different openings in the lubrication system.

To get accurate results from the test, make sure you follow the manufacturer's recommendations and compare your findings to specifications. Low oil pressure readings can be caused by internal component wear, pump-related problems, low oil level, contaminated oil, or low oil viscosity. An overfilled crankcase, high oil viscosity, or a faulty pressure regulator can cause high oil pressure readings.

Stethoscope

Some engine sounds can be easily heard without using a listening device, but others are impossible to hear unless they are amplified. A stethoscope is very helpful in locating engine noise by amplifying the sound waves. It can also distinguish between normal and abnormal noise. The procedure for using a stethoscope is simple. Use the metal prod to trace the sound until it reaches its maximum intensity. Once the precise location has been discovered, the sound can be better evaluated. A sounding stick, which is nothing more than a long, hollow tube, works on the same principle, though a stethoscope gives much clearer results.

The best results, however, are obtained with an electronic listening device. With this tool you can tune into the noise. Doing this allows you to eliminate all other noises that might distract or mislead you.

Cooling System Pressure Tester

A cooling system pressure tester contains a hand pump and a pressure gauge. A hose is connected from the hand pump to a special adapter that fits on the radiator filler neck. This tester is used to pressurize the cooling system and check for coolant leaks. Additional adapters are available to connect the tester to the radiator cap. With the tester connected to the radiator cap, the pressure relief action of the cap may be checked.

Coolant Hydrometer

A coolant hydrometer is used to check the amount of antifreeze in the coolant. This tester contains a pickup hose, coolant reservoir, and squeeze bulb. The pickup hose is placed in the radiator coolant. When the squeeze bulb is squeezed and released, coolant is drawn into the reservoir. As coolant enters the reservoir, a pivoted float moves upward with the coolant level. A pointer on the float indicates the freezing point of the coolant on a scale located on the reservoir housing.

Belt Tension Gauge

A belt tension gauge is used to measure drive belt tension. The belt tension gauge is installed over the belt, and the gauge indicates the amount of belt tension.

Service Manuals

Perhaps the most important tools you will use are service manuals. There is no way a technician can remember all of the procedures and specifications needed to correctly repair all vehicles. Therefore, a good technician relies on service manuals and other information sources for this information. Good information plus knowledge allows a technician to fix a problem with the least bit of frustration and at the lowest cost to the customer.

To obtain the correct engine specifications and other information, you must first identify the engine you are working on. The best source for engine identification is the VIN. The engine code can be interpreted through information given in the service manual. The manual may also help you identify the engine through casting numbers and/or markings on the cylinder block or head.

The primary source of repair and specification information for any car, van, or truck is the manufacturer. The manufacturer publishes service manuals each year, for every vehicle built. Because of the enormous amount of information, some manufacturers publish more than one manual per year per car model. They are typically divided into sections based on the major systems of the vehicle. In the case of engines, there is a section for each engine that may be found in the vehicle. Manufacturers' manuals detail all repairs, adjustments, specifications, diagnostic procedures, and special tools required.

Since many technical changes occur on specific vehicles each year, manufacturers' service manuals need to be constantly updated. Updates are published as service bulletins (often referred to as Technical Service Bulletins or TSBs) that show the changes in specifications and repair procedures during the model year. These changes do not appear in the service manual until the next year. The car manufacturer provides these bulletins to dealers and repair facilities on a regular basis.

Service manuals are also published by independent companies rather than by the manufacturers. However, they pay for and get most of their information from the car makers. They contain component information, diagnostic steps, repair procedures, and specifications for several car makes in one book. Information is usually condensed and is more general in nature than the manufacturer's manuals. The condensed format allows for more coverage in less space and therefore is not always specific. They may also contain several years of models as well as several car makes in one book.

Many of the larger parts manufacturers have excellent guides on the various parts they manufacture or supply. They also provide updated service bulletins on their products. Other sources for up-to-date technical information are trade magazines and trade associations.

The same information that is available in service manuals is now commonly found electronically on compact disks (CD-ROMs), digital video disks (DVDs), and the Internet. A single compact disk can hold a quarter million pages of text, eliminating the need for a huge library to contain all of the printed manuals. Using electronics to find information is also easier and quicker. The disks are normally updated quarterly and not only contain the most recent service bulletins but also engineering and field service fixes. DVDs can hold more information than CDs; therefore, fewer disks are needed with systems that use DVDs. The CDs and DVDs are inserted into a computer. All a technician needs to do is enter vehicle information and then move to the appropriate part or system. The appropriate information will then appear on the computer's screen. Online data can be updated instantly and requires no space for physical storage. These systems are easy to use and the information is quickly accessed and displayed. The computer's keyword, mouse, and/or light pen are used to make selections from the screen's menu. Once the information is retrieved, a technician can read it off the screen or print it out and take it to the service bay.

CROSS-REFERENCE GUIDE

NATEF Task	Job Sheet
A.1	1
A.2	2
A.3	3
A.4	4
A.5	2
A.6	2
A.7	2
A.8	5
A.9	6
A.10	7
A.11	8
A.12	9
A.13	10
A.14	11
A.15	12
A.16	13
A.17	14
B.1	15
B.2	16
B.3	17
B.4	18
B.5	19
B.6	20
B.7	19
B.8	21
B.9	22 & 23 & 24
B.10	25
C.1	26
C.2	26
C.3	27 & 28
C.4	28
C.5	28
C.6	29
C.7	30
C.8	27 & 28
D.1	31 & 32
D.2	33

D.3	34 & 35 & 42
D.4	36
D.5	37
D.6	38 & 40
D.7	39
D.8	41
D.9	43
D.10	44
D.11	45
E.1.1	46
E.1.2	46
E.2.1	47
E.2.2	47
E.2.3	47
E.3.1	44
E.3.2	48
E.3.3	48
E.3.4	44
E.4.1	49
E.4.2	49
E.4.3	49
F.1	50
F.2	51
F.3	52
F.4	53
F.5	54
F.6	55
F.7	56
F.8	57

JOB SHEETS

ENGINE PERFORMANCE JOB SHEET 1

Filling Out a Work Order

Name ______________________ Station ______________ Date ______________

NATEF Correlation:

This Job Sheet addresses the following NATEF task:

A.1. Complete work order to include customer information, vehicle identifying information, customer concern, related service history, cause, and correction.

Objective

Upon completion of this job sheet, you will be able to prepare a service work order based on customer input, vehicle information, and service history.

Tools and Materials

An assigned vehicle or the vehicle of your choice

Service work order or computer-based shop management package

Parts and labor guide

Work Order Source: Describe the system used to complete the work order. If a paper repair order is being used, describe the source.

PROCEDURE

1. Prepare the shop management software for entering a new work order or obtain a blank paper work order. Task Completed ☐
2. Enter customer information, including name, address, and phone numbers onto the work order. Task Completed ☐
3. Locate and record the vehicle's VIN. Task Completed ☐
4. Enter the necessary vehicle information, including year, make, model, engine type and size, transmission type, license number, and odometer reading. Task Completed ☐
5. Does the VIN verify that the information about the vehicle is correct? __________
6. Normally, you would interview the customer to identify his or her concerns. However to complete this job sheet, assume the only concern is that the engine lacks power and this is especially noticeable during acceleration. For the purpose of this job sheet, there is a dirty fuel filter that must be replaced. Task Completed ☐

7. The history of service to the vehicle can often help diagnose problems as well as indicate possible premature part failure. Gathering this information from the customer can provide some of this information. For this job sheet assume the vehicle has not had a similar problem and was not recently involved in a collision. Service history is further obtained by searching files based on customer name, VIN, and license number. Check the files for any related service work. Task Completed ☐

8. Search for technical service bulletins on this vehicle that may relate to the customer's concern. Task Completed ☐

9. Based on the customer's concern, service history, TSBs, and your knowledge, what is the likely cause of this concern?

____

____

10. Enter this information onto the work order. Task Completed ☐

11. Prepare to make a repair cost estimate for the customer. Identify all parts that may need to be replaced to correct the concern. List these here.

____

____

12. Describe the task(s) that will be necessary to replace the part.

____

____

13. Using the parts and labor guide, locate the cost of the parts that will be replaced and enter the cost of each item onto the work order at the appropriate place for creating an estimate. Task Completed ☐

14. Now, locate the flat rate time for work required to correct the concern. List each task and with its flat rate time.

____

____

15. Multiply the time for each task by the shop's hourly rate and enter the cost of each item onto the work order at the appropriate place for creating an estimate. Task Completed ☐

16. Many shops have a standard amount they charge each customer for shop supplies and waste disposal. For this job sheet, use an amount of ten dollars for shop supplies. Task Completed ☐

17. Add the total costs and insert the sum as the subtotal of the estimate. Task Completed ☐

18. Taxes must be included in the estimate. What is the sales tax rate and does it apply to both parts and labor, or just one of these?

____

____

19. Enter the appropriate amount of taxes to the estimate, than add this to the subtotal. The end result is the estimate to give the customer. Task Completed ☐

20. By law, how accurate must your estimate be?

____

____

21. Generally speaking, the work order is complete and is ready for the customer's signature. However, some businesses require additional information; make sure you enter that information to the work order. On the work order there is a legal statement that defines what the customer is agreeing to. Briefly describe the contents of that statement.

____

____

Problems Encountered

____

____

____

Instructor's Comments

____

____

____

ENGINE PERFORMANCE JOB SHEET 2

Verifying the Condition of and Inspecting an Engine

Name ______________________ Station ______________ Date ______________

NATEF Correlation

This Job Sheet addresses the following NATEF tasks:

A.2. Identify and interpret engine performance concern; determine necessary action.

A.5. Inspect engine assembly for fuel, oil, coolant, and other leaks; determine necessary action.

A.6. Diagnose abnormal engine noise or vibration concerns; determine necessary action.

A.7. Diagnose abnormal exhaust color, odor, and sound; determine necessary action.

Objective

Upon completion of this job sheet, you will be able to identify and interpret engine concerns; inspect an engine assembly for fuel, oil, coolant, and other leaks; diagnose engine noises and vibrations; and diagnose the cause of excessive oil consumption, unusual engine exhaust color, odor, and sounds.

Tools and Materials

A road worthy car

Stethoscope

Protective Clothing

Goggles or safety glasses with side shields

Describe the vehicle being worked on:

Year ________ Make ____________________ Model ____________________

VIN ____________________ Engine type and size ____________________

PROCEDURE

Verify Engine Condition

1. Verifying the customer's concern is typically the first step you should take when diagnosing a problem. If the owner of the vehicle stated a concern, describe it. If there are no customer complaints, describe the general running condition to the best of your knowledge. The concern may be one of performance, smoke, leaks, or noises. In your answer, be sure to completely describe the condition and state when, where, and how the condition occurs.

____

____

____

____

2. Verify the complaint. Describe what you will do in an attempt to duplicate the concern. If necessary, road test the vehicle under the same conditions that are present when the problem normally occurs. Include in this description conditions that will explain when, where, and how the concern occurs.

____

____

____

____

3. Often a customer may only notice poor performance during one condition, whereas the problem may exist at others. Also, observing the performance of an engine during a variety of modes of operation may let you know what is working fine and what does not need to be tested further. Describe how the engine performed during the following conditions:

   A. Starting:

   ____

   B. Idling:

   ____

   C. Slow acceleration:

   ____

   D. Slow cruise:

   ____

   E. Slow deceleration:

   ____

   F. Heavy acceleration:

   ____

   G. Highway cruise:

   ____

   H. Fast deceleration:

   ____

   I. Shut down:

   ____

4. Based on the results of the above checks, what engine systems do you think should be checked to find the cause of the customer's complaint?

____

____

____

Engine Leak Diagnosis

1. After you have completed the above exercise, park the vehicle in a place where the floor is fairly clean and dry. Shut the engine off and allow it to cool. Once the engine is cool, look on the shop floor for any evidence of leaks. Remember, engine leaks can cause safety, driveability, and durability concerns. Describe the location and color of the fluid that leaked on the floor.

____

____

____

WARNING: *Gasoline fumes are extremely dangerous! If ignited, they will cause a very serious explosion and fire, resulting in personal injury and property damage. If you suspect that the leak is a fuel leak, immediately inform your instructor of the problem and take all precautions to avoid igniting the fuel.*

2. Engine fuel leaks are expensive and dangerous, and they should be corrected immediately when they are detected. If gasoline odor occurs inside or near a vehicle, it should be inspected for fuel leaks immediately. Inspect the following parts of the vehicle to locate the source of the fuel leak. Describe what you found at each of these components:

   A. Fuel tank:

   ____

   B. Fuel tank filler cap:

   ____

   C. Fuel lines and filter:

   ____

   D. Mechanical fuel pump (carbureted engines):

   ____

   E. Vapor recovery system lines:

   ____

   F. Carburetor:

   ____

   G. Pressure regulator, fuel rail, and injectors (fuel injected engines):

   ____

3. Based on the above inspection, what do you recommend should be done to the vehicle to correct the fuel leak?

____

____

4. Engine oil leaks may cause an engine to run out of oil, resulting in serious engine damage. Sometimes the leak goes unnoticed, and yet the engine's oil needs to be added to more often than normal. A careful inspection may help you locate the source of the leak; however, if it is difficult to locate the exact cause of an oil leak, the engine should be cleaned first. Carefully look at the following engine parts and describe what you see. If oil is found on or around any of these parts, look carefully at the component directly above the leak. Oil runs down the engine surfaces. Also, if an oil leak is found, check the PCV system for blockage and proper operation. A bad PCV system will allow crankcase pressure to build and this may be the ultimate cause of the oil leak.

   A. Rear main bearing seal:

   ____

   B. Expansion plug in rear camshaft bearing:

   ____

   C. Rear oil gallery plug:

   ____

   D. Oil pan:

   ____

   E. Oil filter:

   ____

   F. Rocker arm covers:

   ____

   G. Intake manifold front and rear gaskets (V-type engines):

   ____

   H. Mechanical fuel pump gasket or worn fuel pump pivot pin:

   ____

   I. Timing gear cover or seal:

   ____

   J. Front main bearing:

   ____

   K. Oil pressure sending unit:

   ____

   L. Distributor O-ring or gasket:

   ____

   M. Engine casting porosity:

   ____

   N. Oil cooler or lines (where applicable):

   ____

5. Based on the above inspection, what do you recommend should be done to the vehicle to correct the oil leak?

____

____

6. If an engine uses excessive oil and there is no evidence of leaks, the oil may be burning in the combustion chambers. If excessive amounts of oil are burned in the combustion chambers, the exhaust contains blue smoke, and the spark plugs may be fouled with oil. Excessive oil burning in the combustion chambers may be caused by worn rings and cylinders or worn valve guides and valve seals. Remove the spark plugs and describe the condition of each.

#1 ______________________________

#2 ______________________________

#3 ______________________________

#4 ______________________________

#5 ______________________________

#6 ______________________________

#7 ______________________________

#8 ______________________________

7. When an engine coolant leak causes low coolant level, the engine quickly overheats and severe engine damage may occur. When you suspect a coolant leak, check the following components and describe what you found:

A. Upper radiator hose:

____

B. Lower radiator hose:

____

C. Heater hoses:

____

D. By-pass hose:

____

E. Water pump:

____

F. Engine expansion plugs or block heater:

____

G. Radiator:

____

H. Thermostat housing:

____

I. Heater core:

____

8. Based on the above inspection, what do you recommend should be done to the vehicle to correct the oil leak?

____

____

9. Some coolant leaks are internal to the engine and will not be visible. A whitish exhaust may be indicative of this sort of coolant leak if it is caused by a leaking head gasket or cracked engine components. Take a look at the exhaust while the engine is idling and describe it:

____

____

10. A cooling system pressure tester is commonly used to locate leaks. The use of this tester is covered in another job sheet in this manual. What page is the procedure found on?

____

Engine Exhaust Diagnosis

1. Some engine problems may be diagnosed by the color, smell, or sound of the exhaust. If the engine is operating normally, the exhaust should be colorless. In severely cold weather, it is normal to see a swirl of white vapor coming from the tailpipe, especially when the engine and exhaust system are cold. This vapor is moisture in the exhaust, which is a normal by-product of the combustion process. Carefully observe the exhaust from your test vehicle and describe what you see and/or smell during the following operating conditions:

   A. Cold start-up:

   ____

   B. Cold idle:

   ____

   C. Warm start-up:

   ____

   D. Warm idle:

   ____

   E. Snap-throttle open:

   ____

   F. Snap-throttle closed:

   ____

2. Based on the above, what are your conclusions? (Use the explanations below to guide your thoughts.)

____

____

If the exhaust is blue, excessive amounts of oil are entering the combustion chamber, and this oil is being burned with the fuel. When the blue smoke in the exhaust is more noticeable on deceleration, the oil is likely getting past the rings into the cylinder. If the blue smoke appears in the exhaust immediately after a hot engine is restarted, the oil is likely leaking down the valve guides.

- If black smoke appears in the exhaust, the air-fuel mixture is too rich. A restriction in the air intake, such as a plugged air filter, may be responsible for a rich air-fuel mixture.
- Gray smoke in the exhaust may be caused by coolant leaking into the combustion chambers. This may be most noticeable when the engine is first started after it has been shut off for over half an hour.
- On catalytic converter-equipped vehicles, a strong sulfur smell in the exhaust indicates a rich air-fuel mixture. Some sulfur smell on these engines is normal, especially during engine warm-up.

3. You may have noticed a change in sound during the above test. If you did, describe the sound change and operating mode in which the sound changed.

____

____

____

____

4. Use these guidelines to determine the possible cause of the sound. Then state your best guess for the cause of the noise.

- When the engine is idling, the exhaust at the tailpipe should have a smooth, even sound.
- If, during idle, the exhaust has a "puff" sound at regular intervals, a cylinder may be misfiring.
- When this sound is present, check the engine's ignition and fuel systems, and the engine's compression.
- If the vehicle has excessive exhaust noise, while the engine is accelerated, check the exhaust system for leaks.
- A small exhaust leak may cause a whistling noise when the engine is accelerated.
- If the exhaust system produces a rattling noise when the engine is accelerated, check the muffler and catalytic converter for loose internal components.
- When the engine has a wheezing noise at idle or with the engine running at higher rpm, check for a restricted exhaust system.

____

____

____

Engine Noise Diagnosis

1. Sounds from the engine itself can help you locate engine problems or help you identify a weakness in the engine before it becomes a big problem. Long before a serious engine failure occurs, there are usually warning noises from the engine. Engine defects such as damaged pistons, worn rings, loose piston pins, worn crankshaft bearings, worn camshaft lobes, and loose and worn valve train components usually produce their own peculiar noises. Certain engine defects also cause a noise under specific engine operating conditions. Since it is sometimes difficult to determine the exact location of an engine noise, a stethoscope may be useful. The stethoscope probe is placed on, or near, the suspected component, and the ends of the stethoscope are installed in your ears. The stethoscope amplifies sound to assist in noise location. When the stethoscope probe is moved closer to the source of the noise, the sound is louder in your ears. If a stethoscope is not available, what can be safely used to amplify the sound and help locate the source of the noise? ____________________

CAUTION: *When placing a stethoscope probe in various locations on a running engine, be careful not to catch the probe or your hands in moving components such as cooling fan blades and belts.*

2. Since a lack of lubrication is a common cause of engine noise, always check the engine oil level and condition prior to noise diagnosis. Carefully observe the oil for contamination by coolant or gasoline. Check the oil level and condition on your test vehicle, then record your findings:

____

____

3. During the diagnosis of engine noises, always operate the engine under the same conditions as those that are present when the noise ordinarily occurs. Remember that aluminum engine components such as pistons expand more when heated than cast iron alloy components do. Therefore, a noise caused by a piston defect may occur when the engine is cold but disappear when the engine reaches normal operating temperature. If the customer has an engine noise concern, describe the noise and state when it occurs.

____

____

____

4. Duplicate the condition at which the noise typically occurs and describe all that you hear as you listen to the engine.

____

____

5. If you verified the customer's concern, use a stethoscope to find the spot where the noise is the loudest. Describe where that is:

____

____

6. What could be causing the noise to be loud at that spot?

____

____

7. To help you understand and use noise as a diagnostic tool, you will be given a description of an engine noise. Using your knowledge and any resources (such as your text book) you have handy, identify the conditions or problems that would cause each of the following noises:

   A. A hollow, rapping noise that is most noticeable on acceleration with the engine cold. The noise may disappear when the engine reaches normal operating temperature.

   ____

   B. A heavy thumping knock for a brief time when the engine is first started after it has been shut off for several hours. This noise may also be noticeable on hard acceleration.

   ____

   C. A sharp, metallic, rapping noise that occurs with the engine idling.

   ____

   D. A thumping noise at the back of the engine.

   ____

   E. A rumbling or thumping noise at the front of the engine, possibly accompanied by engine vibrations. When the engine is accelerated under load, the noise is more noticeable.

   ____

   F. A light, rapping noise at speeds above 35 mph (21 kph). The noise may vary from a light to a heavier rapping sound depending on the severity of the condition. If the condition is very bad, the noise may be evident when the engine is idling.

   ____

   G. A high-pitched, clicking noise is noticeable in the upper cylinder area during acceleration.

   ____

   H. A heavy clicking noise is heard with the engine running at 2,000 to 3,000 rpm. When the condition is severe, a continuous, heavy clicking noise is evident at idle speed.

   ____

   I. A whirring and light rattling noise when the engine is accelerated and decelerated. Severe cases may cause these noises at idle speed.

   ____

   J. A light clicking noise with the engine idling. This noise is slower than piston or connecting rod noise and is less noticeable when the engine is accelerated.

   ____

K. A high-pitched clicking noise that intensifies when the engine is accelerated.

____

L. A noise that is similar to marbles rattling inside a metal can. This noise usually occurs when the engine is accelerated.

____

Problems Encountered

____

____

____

Instructor's Comments

____

____

____

ENGINE PERFORMANCE JOB SHEET 3

Gathering Vehicle Information

Name ______________________ Station ______________ Date ______________

NATEF Correlation

This Job Sheet addresses the following NATEF task:

A.3. Research applicable vehicle and service information, such as engine management system operation, vehicle service history, service precautions, and technical service bulletins.

Objective

Upon completion of this job sheet, you will be able to gather service information about a vehicle and its engine and related systems.

Tools and Materials

Appropriate service manuals

Computer

Protective Clothing

Goggles or safety glasses with side shields

Describe the vehicle being worked on:

Year ________ Make __________________ Model __________________

VIN ____________________

PROCEDURE

1. Using the service manual or other information source, describe what each letter and number in the VIN for this vehicle represents.

____

____

2. While looking in the engine compartment, did you find a label regarding the specifications for the engine? Describe where you found it.

____

____

____

____

3. Summarize the information contained on this label.

4. Using a service manual or electronic database, locate the information about the vehicle's engine. List the major components of the system and describe the primary characteristics of the engine (number of valves, shape/configuration, firing order. etc.).

5. Describe the engine's support systems (fuel, air, ignition, exhaust, and emissions) and their control system(s).

6. Using a service manual or electronic database, locate and record all service precautions regarding working on the engine and its systems as noted by the manufacturer.

7. Using the information that is available, locate and record the vehicle's service history.

8. Using the information sources that are available, summarize all Technical Service Bulletins for this vehicle that relate to the engine and its systems.

____

____

____

____

Problems Encountered

____

____

____

Instructor's Comments

____

____

____

ENGINE PERFORMANCE JOB SHEET 4

Interpreting Identification Numbers

Name ______________________ Station ______________ Date ______________

NATEF Correlation

This Job Sheet addresses the following NATEF task:

A.4. Locate and interpret vehicle and major component identification numbers (VIN, vehicle certification labels, and calibration decals).

Objective

Upon completion of this job sheet, you will be able to locate and interpret vehicle and major component identification numbers (VIN, vehicle certification labels, and calibration decals).

Tools and Materials

Service manual

Describe the vehicle being worked on:

Year ________ Make ____________________ Model ____________________

VIN ____________________ Engine type and size ____________________

PROCEDURE

1. Locate the VIN plate. Where is it?

   ____

2. What code is used to designate the engine size?

   ____

3. What is the code used in this VIN to designate model year?

   ____

4. What is the production number of this vehicle?

   ____

5. Look under the hood and find the VECI. Where is it?

   ____

   ____

6. Does the VECI contain a calibration number? If not, is there a separate calibration label? Where?

   ____

   ____

7. Locate the label on the vehicle that states what the recommended tire inflation is. Where is it and what is the recommendation?

____

____

8. Locate the label that states what color paint the vehicle has. Where is it and what color is designated?

____

____

9. How do you know what type of restraint system the vehicle is equipped with?

____

____

Problems Encountered

____

____

____

Instructor's Comments

____

____

____

ENGINE PERFORMANCE JOB SHEET 5

Using a Vacuum Gauge

Name ______________________ Station ______________ Date ______________

NATEF Correlation

This Job Sheet addresses the following NATEF task:

A.8. Perform engine absolute (vacuum/boost) manifold pressure tests; determine necessary action.

Objective

Upon completion of this job sheet, you will be able to properly perform engine absolute (vacuum/boost) manifold pressure tests.

Tools and Materials

Vacuum gauge

Various lengths of vacuum hose

Tee hose fittings

Protective Clothing

Safety glasses

Describe the vehicle being worked on:

Year ________ Make ____________________ Model ____________________

VIN ____________________ Engine type and size ____________________

Describe the general operating condition: ______________________________

____

Give a definition of vacuum ______________________________________

PROCEDURE

1. Carefully look over the engine's intake manifold to identify vacuum hoses. Select one that is small and easily accessible. DO NOT disconnect it until you have the approval of your instructor. Also make sure the hose is below the throttle plate assembly. Task Completed ☐
2. With the engine off, disconnect the selected hose. Connect the Tee fitting to the end of the hose and connect another hose from the Tee fitting to the place where the vacuum hose was originally connected. Task Completed ☐
3. Connect the vacuum gauge to the remaining connection at the Tee fitting. Task Completed ☐
4. Start the engine and observe the vacuum gauge. Task Completed ☐

5. Describe the gauge reading and the action of the needle.

____

____

6. Quickly open the throttle and allow it to quickly close. Task Completed ☐

7. Describe the gauge reading and the action of the needle.

____

____

8. Turn off the engine and disconnect the vacuum gauge and hoses. Then reconnect the engine's hose to its appropriate connector. Task Completed ☐

Summarize what you observed.

____

____

____

Problems Encountered

____

____

____

Instructor's Comments

____

____

____

ENGINE PERFORMANCE JOB SHEET 6

Conduct a Cylinder Power Balance Test

Name ______________________ Station ______________ Date ______________

NATEF Correlation

This Job Sheet addresses the following NATEF task:

A.9. Perform cylinder power balance test; determine necessary action.

Objective

Upon completion of this job sheet, you will be able to conduct a cylinder power balance test and accurately interpret the results.

Tools and Materials

Tune-up scope

Protective Clothing

Goggles or safety glasses with side shields

Describe the vehicle being worked on:

Year ________ Make ____________________ Model ____________________

VIN ____________________ Engine type and size ____________________

PROCEDURE

1. What is the firing order for this engine and how are the cylinders numbered?

   ____

   ____

   ____

2. Describe the general running condition of the engine.

   ____

   ____

   ____

3. Connect the scope to the engine according to the instructions given with the equipment. Task Completed ☐

4. Conduct a cylinder power balance test on the engine, using proper testing procedures (make sure not to short a cylinder for a long time if the vehicle is equipped with a catalytic converter), then record the results below:

Cylinder #	1	2	3	4	5	6	7	8
RPM loss	____	____	____	____	____	____	____	____

5. Describe what is indicated by the results of this test:

____

____

____

6. Briefly explain what is actually being measured by a cylinder power balance test.

____

____

____

7. Based on the test results, describe the condition of the engine. (Does it agree with your original description of the engine?)

____

____

____

8. How and why would the readings be different if the camshaft intake lobes for the number two cylinder were severely worn?

____

____

____

Problems Encountered

____

____

____

Instructor's Comments

____

____

____

ENGINE PERFORMANCE JOB SHEET 7

Perform a Cylinder Cranking Compression Test

Name ______________________ Station ______________ Date ______________

NATEF Correlation

This Job Sheet addresses the following NATEF task:

A.10. Perform cylinder cranking compression test; determine necessary action.

Objective

Upon completion of this job sheet, you will be able to properly perform cylinder compression tests.

Tools and Materials

Spark plug socket, extensions, and ratchet
Compression tester
Fender covers
Air blower
Small oil can with flexible nozzle

Protective Clothing

Safety goggles or glasses with side shields

Describe the vehicle being worked on:

Year ________ Make __________________ Model __________________

VIN __________________ Engine type and size __________________

Describe the general operating condition: ______________________________

____

PROCEDURE

1. Disable the ignition or fuel injection system. Task Completed ☐
2. Remove the spark plug cables from all spark plugs. Use an air blower to clean around the spark plugs. Remove all of the spark plugs and place them on a clean surface in the order in which they are removed from the engine. Task Completed ☐
3. Remove the air cleaner from the car. Block the throttle in its wide-open position. Task Completed ☐
4. Install the compression testing adapter (if necessary) into the number 1 spark plug hole, and connect the compression tester to it. Task Completed ☐
5. Crank the engine through four compression strokes and record the reading from the compression gauge on the chart at the end of this job sheet. Task Completed ☐
6. Remove the compression adapter from the spark plug hole. Squirt a small amount of oil into the cylinder through the spark plug hole. Task Completed ☐

7. Repeat steps 4 and 5 on the same cylinder. Task Completed ☐
8. Complete steps 4, 5, 6, and 7 for each of the remaining cylinders in the engine. Task Completed ☐
9. Clean each spark plug and replace each spark plug in the cylinder from where it was removed. Reconnect the spark plug cables. Remove whatever tool is blocking the throttle plate open, and reinstall the air cleaner. Task Completed ☐
10. Start the engine and check its operation to determine if everything was connected back properly. Task Completed ☐

Problems Encountered

____

____

____

Instructor's Comments

____

____

____

REPORT SHEET ON COMPRESSION TEST

Compression for Each Cylinder Cylinder No.	*Dry*	*Wet*
1.		
2.		
3.		
4.		
5.		
6.		
7.		
8.		

Conclusions and Recommendations ____

____

____

ENGINE PERFORMANCE JOB SHEET 8

Perform an Engine Running Compression Test

Name ______________________ Station ______________ Date ______________

NATEF Correlation:

This Job Sheet addresses the following NATEF task:

A.11. Perform engine running compression test; determine necessary action.

Objective

Upon completion of this job sheet, you will be able to perform an engine running compression test, evaluate the condition of an engine, and determine the services that are required.

Tools and Materials

Service manual

Jumper wire

Compression tester

Protective Clothing

Goggles or safety glasses with side shields

Describe the vehicle being worked on:

Year __________ Make ____________________ Model ____________________

VIN ____________________ Engine type and size ____________________

PROCEDURE

1. A running compression test can be done on all cylinders or the one or ones that have been identified as having a problem. Have you identified the bad cylinders in this engine? If so, what test(s) did you use to identify it?

   ____

   ____

2. Briefly explain the advantage a running compression test has over a cranking compression test.

   ____

   ____

3. Start the engine to allow it to slightly warm up. Make sure it does not get too hot to safely work on it. Task Completed ☐

4. Remove one spark plug (test the bad cylinder, if you know where it is). Ground the spark plug wire with a jumper wire to prevent module damage. Check to make sure there is only spark plug for that cylinder. If there is a second one, remove and ground the spark wire to it. What cylinder are you checking?

____

____

5. Disconnect the injector to that cylinder on a port-type fuel injection system. Task Completed ☐
6. Choose the correct compression hose adapter for the compression tester. Task Completed ☐
7. Make sure the seal is on the end of the hose and tightly install the hose into the spark plug bore. Task Completed ☐
8. Attach the compression gauge to the end of the hose. Task Completed ☐
9. Start the engine and allow it to idle. Watch the reading on the compression gauge. What was the reading? If you failed to get an accurate reading, relieve the pressure in the gauge and check the pressure again.

____

____

10. Using the throttle linkage, quickly open and close the throttle while observing the compression gauge. What was your reading?

____

____

11. Turn off the engine. Task Completed ☐
12. What are your conclusions from this test? Unless the service manual tells you otherwise, compression pressure at idle should be about half of cranking compression pressure or 50-75 psi. The pressure should increase to about 80% of cranking pressure (80 to 120 psi) when the throttle is snapped open.

____

____

13. Repeat the steps on all cylinders you desire to test. Record the results of the tests.

____

____

14. Was there more than a 20-psi difference between cylinders?

____

____

15. What do you know about this engine after you have completed the test?

____

____

Problems Encountered

____

____

____

Instructor's Comments

____

____

____

ENGINE PERFORMANCE JOB SHEET 9

Perform a Cylinder Leakage Test

Name ______________________ Station ______________ Date ______________

NATEF Correlation

This Job Sheet addresses the following NATEF task:

A.12. Perform cylinder leakage test; determine necessary action.

Objective

Upon completion of this job sheet, you will be able to properly perform cylinder leakage tests.

Tools and Materials

Breakover bar
Chalk
Compressed air
Fender covers
Indicator light
Jumper lead
Leakage tester and whistle
Radiator coolant (if applicable)
Screwdrivers
Service manual
Socket set
Spark plug socket and ratchet
TDC indicator
Test adapter hose

Protective Clothing

Safety goggles or glasses with side shields

Describe the vehicle being worked on:

Year __________ Make ______________________ Model ______________________

VIN ______________________ Engine type and size ______________________

Describe general condition: ____

____

PROCEDURE

CAUTION: *Very high voltages are present with high-energy ignition systems. Do NOT use this procedure on cars with Distributorless ignition systems unless the ignition system is totally disabled.*

1. Check the coolant level and fill, if needed. Run the engine until it reaches normal operating temperature. Then, turn it off and disable the ignition or fuel injection system. Task Completed ☐

2. Disconnect the spark plug cables from the plugs. Use compressed air to clean all foreign matter out of the plug wells. Remove all spark plugs. Set the plugs on a workbench or other clean surface in the order in which they were removed. Remove all plug gaskets or tubes from the cylinder head. Task Completed ☐

3. Remove the air cleaner. Block the throttle plate in a wide-open position using a screwdriver or similar tool. Disconnect the PCV hose from the crankcase. Task Completed ☐

4. Install the test adapter hose in the number 1 cylinder spark plug hole. Connect the tester whistle to adapter hose. Task Completed ☐

5. Using a socket and ratchet on the crankshaft pulley nut or bolt, slowly rotate the engine in its normal direction until the test whistle sounds indicating the beginning of the compression stroke. Continue rotation until the timing mark on the crankshaft pulley lines up with the engine-timing pointer on the timing chain cover. Remove the tester whistle from the adapter hose. Task Completed ☐

6. Connect the tester to the adapter hose. Does the gauge show more than 20 percent leakage? If so, look for air leaking from the throttle plate, tailpipe, or crankcase and for air bubbles in the radiator. Record your results on the Report Sheet on Cylinder Leakage Test found at the end of this job sheet. Task Completed ☐

7. Disconnect the tester from the adapter hose. Resume rotating engine using socket and ratchet on the crankshaft nut or bolt until the next appropriate cylinder mark lines up with the chalk mark on the engine. It may be necessary to use the whistle to locate the TDC. Task Completed ☐

8. Remove the adapter from the previously tested cylinder and install it in the plug hole of the next cylinder in the firing order. Task Completed ☐

9. Repeat steps 6, 7, and 8 on each remaining cylinder. Task Completed ☐

Problems Encountered

____

____

____

Instructor's Comments

____

____

____

Name ______________________ Station ______________ Date ______________

REPORT SHEET ON CYLINDER LEAKAGE TEST

No.	*Percentage*	*Leakage From*
1.		
2.		
3.		
4.		
5.		
6.		
7.		
8.		

Conclusions and Recommendations ____

____

____

ENGINE PERFORMANCE JOB SHEET 10

Scope Testing an Ignition System

Name ______________________ Station ______________ Date ______________

NATEF Correlation

This Job Sheet addresses the following NATEF task:

A.13. Diagnose engine mechanical, electrical, electronic, fuel, and ignition concerns with an oscilloscope and engine diagnostic equipment; determine necessary action.

Objective

Upon completion of this job sheet, you will be able to diagnose an engine's mechanical, electrical, electronic, fuel, and ignition concerns with an oscilloscope and engine diagnostic equipment.

Tools and Materials

Service manual for the vehicle below

An engine diagnostic scope

Protective Clothing

Goggles or safety glasses with side shields

Describe the vehicle being worked on:

Year ________ Make __________________ Model __________________

VIN ____________________ Engine type and size ____________________

Firing order ____________________

Describe general operating condition:

____

PROCEDURE

1. Connect the scope leads to the vehicle. Task Completed ☐
2. Set the scope to look at the secondary ignition circuit in the display or parade pattern. Task Completed ☐
3. Start the engine and observe the height of the firing lines. Are they within 3 kV of each other? _______ If not, which cylinders are the most different from the rest? _______ Are the heights of the firing lines all below 11 kV? ______ Are the heights of the firing lines above 8 kV? ______ Based on these findings, what do you conclude?

   ____

   ____

   ____

   ____

4. Switch the scope to show the patterns in raster. Observe the length, height, and shape of the spark line for each cylinder. Describe them.

____

____

____

5. Based on these findings, what do you conclude?

____

____

____

6. Describe the general appearance of the intermediate section of the pattern for each cylinder.

____

____

____

7. Based on these findings, what do you conclude?

____

____

____

8. Describe the general appearance of the dwell section for each cylinder.

____

____

____

9. Based on these findings, what do you conclude?

____

____

____

10. Based on the appearance of all sections of the scope pattern for each cylinder, what are your recommendations and conclusions?

____

____

____

Problems Encountered

Instructor's Comments

ENGINE PERFORMANCE JOB SHEET 11

Using an Exhaust Gas Analyzer

Name ______________________ Station ______________ Date ______________

NATEF Correlation

This Job Sheet addresses the following NATEF task:

A.14. Prepare 4 or 5 gas analyzer; inspect and prepare vehicle for test, and obtain exhaust readings; interpret readings, and determine necessary action.

Objective

Upon completion of this job sheet, you will be able to prepare a 4 or 5 gas analyzer, inspect and prepare a vehicle for test, obtain exhaust readings, as well as interpret the readings.

Tools and Materials

An exhaust gas analyzer

Protective Clothing

Goggles or safety glasses with side shields

Describe the vehicle being worked on:

Year ________ Make ____________________ Model ____________________

VIN ____________________ Engine type and size ____________________

Describe general operating condition:

____

Type of gas analyzer used: ______________________________________

Check the gases that the tool will measure: HC ____ CO ____ CO_2 ____ O_2 ____ NO_x ____

In what units of measurement is each of these gases measured?

HC ______________ CO ______________ CO_2 ______________

O_2 ______________ NO_x ______________

Check with your instructor for this information—The maximum acceptable readings for this year and model vehicle are:

HC ______________ CO ______________ CO_2 ______________

O_2 ______________ NO_x ______________

PROCEDURE

1. Turn on the exhaust gas analyzer and allow it to warm up. Task Completed ☐

2. Insert the tester's exhaust probe into the exhaust pipe of the vehicle. Task Completed ☐

3. Install the shop's exhaust ventilation tubing into the vehicle's exhaust pipe. Task Completed ☐

4. Make sure the meters on the exhaust gas analyzer are properly calibrated and adjusted. Task Completed ☐

5. Block the wheels of the vehicle and set the parking brake. Task Completed ☐

6. Position the analyzer so that it is easily read from the driver's seat of the vehicle. Task Completed ☐

   NOTE: *It takes a few seconds for the analyzer to measure the exhaust gases, so watch the analyzer after the event not during it!*

7. Start the engine. Task Completed ☐

8. Record the exhaust readings from starting.

   HC ______________ CO ______________ CO_2 ______________

   O_2 ______________ NO_x ______________

9. After start-up, the exhaust readings will settle at some point for a very short period of time. Record those readings.

   HC ______________ CO ______________ CO_2 ______________

   O_2 ______________ NO_x ______________

10. Once the engine has been warmed to normal operating conditions, the exhaust readings will stabilize. What are the readings with the engine warmed up?

    HC ______________ CO ______________ CO_2 ______________

    O_2 ______________ NO_x ______________

11. Turn off the engine and put away the exhaust gas analyzer. Task Completed ☐

Describe what you observed and explain to the best of your knowledge why the readings changed.

____

____

____

Problems Encountered

____

____

____

Instructor's Comments

____

____

____

ENGINE PERFORMANCE JOB SHEET 12

Engine Temperature Check

Name ______________________ Station ______________ Date ______________

NATEF Correlation

This Job Sheet addresses the following NATEF task:

A.15. Verify engine operating temperature; determine necessary action.

Objective

Upon completion of this job sheet, you will be able to verify the operating temperature of an engine and diagnose the cause of abnormal temperatures.

Tools and Materials

Thermometer

Duct tape

Pyrometer

Protective Clothing

Goggles or safety glasses with side shields

Describe the vehicle being worked on:

Year ________ Make ____________________ Model ____________________

VIN ____________________ Engine type and size ____________________

PROCEDURE

1. Engines are designed to run within a narrow range of temperatures. When they operate within that range, they are running efficiently. Typically, abnormal temperatures are only noticed by the owner when there is insufficient heat inside the vehicle during cold weather or when steam rolls out from under the hood. Why would an engine that is running at lower than normal temperatures be less efficient?

   ____

   ____

   ____

   ____

2. Name two things that may cause an engine to operate at lower than normal temperatures.

   ____

   ____

3. Name the part of the cooling system that has the responsibility for maintaining and regulating the temperature of the coolant.

____

4. Higher than normal operating temperatures will also affect the engine's efficiency. Name five problems that may cause the engine to run hot.

____

____

____

____

____

5. There are times when the engine seems to be running colder or hotter than normal but the temperature gauge reads normally. In these cases, the engine's coolant temperature should be measured and compared to the reading on the vehicle's temperature gauge. The temperature of an engine should also be verified when the activity of the powertrain computer seems to indicate that it is responding to a temperature that is different than what the temperature gauge and/or engine coolant temperature sensor indicates. Engine temperature is commonly measured with a pyrometer or a thermometer. Before measuring, look up the temperature specification for the engine's thermostat. The specification is: _______________

   Source of the specification was: _______________

6. Tape the temperature-sensing bulb of the thermometer to the upper radiator hose. Task Completed ☐

7. Operate the engine for 15 minutes at idle speed. Task Completed ☐

8. Now read the temperature indicated on the thermometer. The reading is:

   ____

9. Compare the reading with the specification and state the difference. Explain why it may be different.

____

____

____

____

Problems Encountered

___

___

___

Instructor's Comments

___

___

___

ENGINE PERFORMANCE JOB SHEET 13

Test Cooling System

Name ______________________ Station ______________ Date ______________

NATEF Correlation

This Job Sheet addresses the following NATEF task:

A.16. Perform cooling system pressure tests; check coolant condition; inspect and test radiator, pressure cap, coolant recovery tank, and hoses; perform necessary action.

Objective

Upon completion of this job sheet, you will be able to perform system pressure tests, check coolant condition, inspect and test the radiator, pressure cap, coolant recovery tank, and hoses.

Tools and Materials

Clean cloth

Cooling system tester

Wire

Protective Clothing

Goggles or safety glasses with side shields

Describe the vehicle being worked on:

Year ________ Make __________________ Model __________________

VIN __________________ Engine type and size __________________

Describe general operating condition:

____

PROCEDURE

1. Wipe off the radiator filler neck and inspect it. Task Completed ☐

   Is the sealing seat free of accumulated dirt, nicks, or anything that might prevent a good seal? ☐ Yes ☐ No

   Are the cams on the out-turned flange bent or worn? ☐ Yes ☐ No

2. Inspect the tube from the radiator to the expansion tank for dents and other obstructions. Run wire through the tube to be certain it is clear. Task Completed ☐

3. To test the cooling system for external leaks, attach a cooling system tester to the appropriate flexible adapter and carefully pump up pressure. Looking at tester's gauge, bring the pressure up to the proper cooling system test point indicated on the dial face. Task Completed ☐

Does the pressure begin to drop? ☐ Yes ☐ No

If so, check all external connections, including hoses, gaskets, and heater core for leaks. Task Completed ☐

4. Check all points closely for small pinhole leaks or potentially dangerous weak points in the system. Task Completed ☐

To test for internal leaks:

5. Remove the flexible adapter from radiator, and replace the pressure cap. Task Completed ☐
6. Start the engine and allow it to reach normal operating temperature so the thermostat will open fully. Task Completed ☐
7. Slowly and carefully remove the pressure cap and replace it with the flexible adapter. Task Completed ☐

WARNING: *Engine coolant will be under pressure. Always wear hand protection and safety goggles or glasses with side shields when doing this task.*

8. Lock the tester slowly into place. Watch for pressure buildup. Task Completed ☐

WARNING: *If the pressure builds up suddenly due to an internal leak, remove the tester immediately.*

9. If no immediate pressure buildup is visible on the gauge, keep the engine running. Pressurize the cooling system. Task Completed ☐
10. Does the gauge fluctuate? If so, the cooling system has a compression or combustion leak. ☐ Yes ☐ No
11. Before removing it, hold the tester body and press the pressure release button against some object on the car to relieve the pressure in the system. Task Completed ☐

WARNING: *Always wear hand protection and safety goggles or glasses with side shields when doing this task. Avoid the hot coolant and steam.*

12. After the pressure has been relieved, shield your hands with a cloth wrapped around the flexible adapter and filler neck. Slowly turn the adapter cap from the lock position to the safety unlock position of the radiator filler neck cam. Do not detach the flexible adapter from the filler neck. Allow the pressure to dissipate completely in this safety position. Then you can remove the adapter and reinstall the radiator cap. Task Completed ☐

Problems Encountered

____

____

____

Instructor's Comments

____

____

____

ENGINE PERFORMANCE JOB SHEET 14

Valve Timing Check

Name ______________________ Station ______________ Date ______________

NATEF Correlation

This Job Sheet addresses the following NATEF task:

A.17. Verify correct camshaft timing.

Objective

Upon completion of this job sheet, you will be able to verify camshaft timing according to the manufacturer's specifications.

Tools and Materials

Spark plug socket
Compression gauge
Remote starter switch
Flashlight
Miscellaneous hand tools
Large breaker bar
New rocker arm or camshaft cover gasket

Protective Clothing

Goggles or safety glasses with side shields

Describe the vehicle being worked on:

Year ________ Make ________________ Model ________________

VIN ________________ Engine type and size ________________

PROCEDURE

1. If the timing belt or chain has slipped on the camshaft sprocket, the engine may fail to start because the valves are not properly timed in relation to the crankshaft. When the timing belt or chain has only slipped a few cogs on the camshaft sprocket, the engine has a lack of power, and fuel consumption is excessive. To check valve timing, begin by removing the spark plug from number 1 cylinder. Task Completed ☐
2. Disconnect the positive primary wire from the ignition coil to disable the ignition system. Task Completed ☐
3. Connect a remote control switch to the starter solenoid terminal and the battery terminal on the solenoid. Task Completed ☐
4. Place your thumb on top of the spark plug hole at cylinder #1. If this hole is not accessible, place a compression gauge in the opening. Task Completed ☐

5. Crank the engine until compression is felt at the spark plug hole. Then, slowly crank the engine until the timing mark lines up with the zero-degree position on the timing indicator. The number 1 piston is now at TDC on the compression stroke. On many engines, the timing mark is on the crankshaft pulley, and the timing indicator is mounted above the pulley. Task Completed ☐

6. Now, slowly crank the engine for one revolution until the timing mark lines up with the zero-degree position on the timing indicator. The number 1 piston is now at TDC on the exhaust stroke. Task Completed ☐

7. Remove the rocker arm or camshaft cover and install a breaker bar and socket on the crankshaft pulley nut. Observe the valve action while rotating the crankshaft about 30 degrees before and after TDC on the exhaust stroke. In this crankshaft position, the exhaust valve should close a few degrees after TDC on the exhaust stroke, and the intake valve should open a few degrees before TDC on the exhaust stroke. Is this what you observed? Task Completed ☐

   ____

8. If the valves did not open properly in relation to the crankshaft position, the valve timing is not correct. What should you do to correct it?

   ____

   ____

   ____

   ____

9. If the timing was correct, reinstall the rocker arm or camshaft cover with a new gasket. Tighten the attaching bolts to the proper specification. The recommended torque is ______________

10. Reinstall the spark plug and tighten it to the proper specification. The recommended torque is ______________

Problems Encountered

____

____

____

Instructor's Comments

____

____

____

ENGINE PERFORMANCE JOB SHEET 15

Retrieve Codes from the Computer of an Engine Control System

Name ____________________ Station ______________ Date ______________

NATEF Correlation

This Job Sheet addresses the following NATEF task:

B.1. Retrieve and record stored OBD I diagnostic trouble codes; clear codes.

Objective

Upon completion of this job sheet, you will be able to retrieve and record OBD I diagnostic trouble codes.

Tools and Materials

A Chrysler product with Flash code capability

Appropriate service manual

Protective Clothing

Goggles or safety glasses with side shields

Describe the vehicle being worked on:

Year ________ Make ____________________ Model ____________________

VIN ____________________ Engine type and size ____________________

Describe general operating condition:

____

PROCEDURE

1. Block the drive wheels. Task Completed ☐
2. Apply the parking brake and start the engine. Task Completed ☐
3. Press on the brake pedal. Move the shift lever through all positions and return it to park. Task Completed ☐
4. Turn the A/C on and off if the vehicle is so equipped. Task Completed ☐
5. Turn the engine off. Task Completed ☐
6. Turn the ignition switch on-off, on-off, and on. Task Completed ☐
7. Count the flashes of the MIL and record the DTCs below.

____

____

8. What is indicated by the flash codes?

____

____

____

Problems Encountered

____

____

____

Instructor's Comments

____

____

____

ENGINE PERFORMANCE JOB SHEET 16

Conduct a Diagnostic Check on an Engine Equipped with OBD II

Name ______________________ Station ______________ Date ______________

NATEF Correlation

This Job Sheet addresses the following NATEF task:

B.2. Retrieve and record stored OBD II diagnostic trouble codes; clear codes, when applicable.

Objective

Upon completion of this job sheet, you will be able to retrieve and record stored OBD II diagnostic trouble codes.

Tools and Materials

A vehicle equipped with OBD II

Scan tool

Service manual

Protective Clothing

Goggles or safety glasses with side shields

Describe the vehicle being worked on:

Year ________ Make ____________________ Model ____________________

VIN ____________________ Engine type and size ____________________

Describe general operating condition:

____

PROCEDURE

1. Describe the scan tool being used:

   Model ____

2. Check all vehicle grounds, including the battery and computer ground, for clean and tight connections. Comments:

   ____

3. Perform a voltage drop test across all related ground circuits. State where you tested and what your findings were:

   ____

   ____

   ____

4. Check all vacuum lines and hoses, as well as the tightness of all attaching and mounting bolts in the induction system. Comments:

____

5. Check for damaged air ducts. Comments:

____

6. Check the ignition circuit, especially the secondary cables for signs of deterioration, insulation cracks, corrosion, and looseness. Comments:

____

7. Are there any unusual noises or odors? If there are, describe them and tell what may be causing them.

____

____

8. Inspect all related wiring and connections at the PCM. Comments:

____

____

9. Gather all pertinent information about the vehicle and the customer's complaint. This should include detailed information about any symptoms from the customer, a review of the vehicle's service history, published TSBs, and the information in the service manual. Task Completed ☐

10. Are there any vacuum leaks? ___ Yes ___ No

11. Is the engine's compression normal? ___ Yes ___ No

12. Is the ignition system operating normally? ___ Yes ___ No

13. Are there any obvious problems with the air-fuel system? ___ Yes ___ No

14. Your conclusions from the above:

____

____

15. Check the operation of the MIL by turning the ignition ON. Describe what happened and what it means.

____

____

16. Connect the scan tool to the DLC. Task Completed ☐

17. Enter the vehicle identification information into the scan tool. Task Completed ☐

18. Then retrieve the DTCs with the scan tool. Task Completed ☐

19. List all codes retrieved by the scan tool.

____

____

____

____

20. Conclusions from these tests and checks:

____

____

____

____

21. After all codes have been recorded and repairs made or planned, the DTCs should be cleared. This is done to verify that the service corrected the problem. Codes can be cleared with the scan tool. Make sure the scan tool is connected to the DLC. Task Completed ☐

22. Turn the ignition switch ON. Task Completed ☐

23. Erase the DTCs and freeze frame data with the OBD II scan tool. Make sure to follow the instructions given on the tool. Task Completed ☐

24. Codes can also be cleared by removing certain fuses in the system. Check the service manual to identify which fuses will clear the codes and state them here.

____

____

25. On some vehicles, the codes can be cleared by disconnecting the negative battery cable for more than one minute. If you do this, the system must be reinitialized. How do you do this?

____

____

Problems Encountered

____

____

____

Instructor's Comments

____

____

____

ENGINE PERFORMANCE JOB SHEET 17

Interpreting Codes from the Computer of an Engine Control System

Name ______________________ Station ______________ Date ______________

NATEF Correlation

This Job Sheet addresses the following NATEF task:

B.3. Diagnose the causes of emissions or driveability concerns resulting from malfunctions in the computerized engine control system with stored diagnostic trouble codes.

Objective

Upon completion of this job sheet, you will be able to diagnose the causes of emissions or driveability concerns resulting from the failure of the computerized engine controls with stored diagnostic trouble codes.

Tools and Materials

Scan tool

Digital multimeter

Service manual

Protective Clothing

Goggles or safety glasses with side shields

Describe the vehicle being worked on:

Year ________ Make ________________ Model ________________

VIN ________________ Engine type and size ________________

PROCEDURE

1. Refer to the listing of codes given in the service manual. Briefly describe how the codes are grouped by letter and number.

   ____

   ____

   ____

   ____

2. Describe what is indicated by the following codes:

   P0037: ______________________________________

   P0143: ______________________________________

   P0010: ______________________________________

P0100: ______________________________

P2195: ______________________________

3. Conduct all preliminary checks of the engine, electrical system, and vacuum lines. Did you find any problems?

____

____

____

4. Connect the scan tool to the DLC. Task Completed ☐

5. Enter the vehicle information into the scan tool. Task Completed ☐

6. Retrieve the DTCs with the scan tool. What are they?

____

____

____

7. Refer to the service manual and locate the description of the DTCs retrieved from the computer. What do they signify?

____

____

____

____

____

____

8. For each of the codes, summarize what steps the manufacturer recommends for locating the exact cause of the problem.

____

____

____

____

____

____

9. Follow those steps and summarize what you found.

____

____

____

____

____

10. After correcting the problem, recheck the DTCs to make sure you properly corrected the problem. Task Completed ☐

Problems Encountered

____

____

____

Instructor's Comments

____

____

____

ENGINE PERFORMANCE JOB SHEET 18

No-Code Diagnostics

Name ______________________ Station ______________ Date ______________

NATEF Correlation

This Job Sheet addresses the following NATEF task:

B.4. Diagnose emissions or driveability concerns resulting from malfunctions in the computerized engine control system with no stored diagnostic trouble codes; determine necessary action.

Objective

Upon completion of this job sheet, you will be able to diagnose emissions or driveability concerns resulting from failure of computerized engine controls with no stored diagnostic codes.

Tools and Materials

Service manual

Digital multimeter

Lab scope

Protective Clothing

Goggles or safety glasses with side shields

Describe the vehicle being worked on:

Year ________ Make ____________________ Model ____________________

VIN ____________________ Engine type and size ____________________

PROCEDURE

1. Determining which part or area of a computerized engine control system is defective requires having a thorough knowledge of how the system works and following a logical troubleshooting process. Electronic engine control problems are usually caused by defective sensors and, to a lesser extent, output devices. The logical procedure in most cases is, therefore, to check the input sensors and wiring first, then the output devices and their wiring, and, finally, the computer. Most late-model computerized engine controls have self-diagnosis capabilities. A malfunction recognized by the computer is stored in the computer's memory as a trouble code. Stored codes can be retrieved and the indicated problem areas checked further. When there is a problem but no codes are stored, the cause of the problem is something not monitored by the computer or is something working within the accept-

able range as seen by the computer. List the components and systems that are monitored by the computer of the vehicle you are working on.

____

____

____

____

2. State the major problem you are diagnosing and describe the general condition of the vehicle.

____

____

3. Diagnostics should begin with a visual inspection. Check the condition of the air filter and related hardware around the filter. Summarize your findings.

____

____

4. Inspect the entire PCV system. Summarize your findings.

____

____

5. Check to make sure the EVAP canister is not saturated or flooded. Summarize your findings.

____

____

6. Check the battery and its cables, and the vehicle's wiring harnesses, connectors, and charging system for loose, corroded, or damaged connections. Summarize your findings.

____

____

7. Check the condition of the battery and its terminals and cables. Summarize your findings.

____

____

8. Make sure all vacuum hoses are connected and are not pinched or cut. Summarize your findings.

____

____

9. Check all sensors and actuators for signs of physical damage. Summarize your findings.

____

____

10. Based on the symptoms, describe the systems and components that you feel could be the cause of the problem if the cause was not discovered in the visual inspection.

____

____

11. Most sensors and output devices can be checked with an ohmmeter. List the ones you suspect as being faulty and that you will test with an ohmmeter.

____

____

12. For those components, list the resistance specifications and your measurements. Then, state your conclusions.

____

____

13. Many sensors, output devices, and circuit wiring can be diagnosed by checking the voltage to and from them. List the ones you suspect as being faulty and that you will test with a voltmeter.

____

____

14. For those components, list the voltage specifications and your measurements. Then, state your conclusions.

____

____

15. The activity of sensors and actuators can be monitored with a lab scope. By watching their activity, you are doing more than testing them. Often problems elsewhere in the system will cause a device to behave abnormally. These situations are identified by the trace on a scope and by the technician's understanding of a scope and the device being monitored. List the ones you suspect as being faulty and that you will test with a lab scope.

____

____

16. Summarize the results of your lab scope checks, then state your conclusions.

____

____

17. If the cause of the problem has not been identified, do a final check on the computer by substituting it with a known good one. Substitution is not an allowable diagnostic method under the mandates of OBD II. Nor is it the most desirable way to diagnose problems. However, to substitute, replace the suspected part with a known good unit and recheck the system. If the system now operates normally, the original part is defective. Task Completed ☐

Problems Encountered

___

___

___

Instructor's Comments

___

___

___

ENGINE PERFORMANCE JOB SHEET 19

Using a Scan Tool on Engine Control Systems

Name ______________________ Station ______________ Date ______________

NATEF Correlation

This Job Sheet addresses the following NATEF tasks:

B.5. Check for module communication (including CAN/BUS systems) errors using a scan tool.

B.7. Obtain and interpret scan tool data.

Objective

Upon completion of this job sheet, you will be able to connect a scan tool to an engine control system and retrieve trouble codes and check for communication errors.

Tools and Materials

Hand tools

Service manual

Scan tool

Protective Clothing

Goggles or safety glasses with side shields

Describe the vehicle being worked on:

Year ________ Make ____________________ Model ____________________

VIN ____________________ Engine type and size ____________________

PROCEDURE

1. Using the service manual as an information source, describe the systems and functions that are controlled by the engine or power train control module.

 ____

 ____

 ____

 ____

2. Check and record the voltage of the battery.

 ____

 ____

3. Is this voltage sufficient for proper operation of the engine control system and its inputs and outputs? Why or why not?

____

____

4. Visually inspect the sensors and switches of the engine control system and summarize your findings.

____

____

5. Check all visible and accessible ground connections for integrity. Describe your findings here.

____

____

6. How would a bad ground affect the operation of a sensor and/or complete circuit?

____

____

7. Check all wiring for signs of burned or chaffed spots, pinched wires, or contact with sharp edges or hot exhaust parts. Describe your findings here.

____

____

8. Visually check all vacuum lines and hoses for integrity. Describe your findings.

____

____

9. Using the service manual as a guide, describe the procedure for retrieving trouble codes from the engine (power train) control module.

____

____

10. Connect the scan tool. Task Completed ☐

11. According to the service manual, what codes would indicate that there is a LAN/CAN/BUS communication problem?

____

____

12. If there is a communication problem, follow the instructions given in the service manual to check the BUS line. Briefly describe them here.

____

____

13. Conduct those checks and describe your findings.

____

____

14. Repair or replace the components that are causing the communication error. Task Completed ☐

15. Clear the codes and recheck for DTCs. Task Completed ☐

16. If the communication problem is solved, inspect and repair the trouble areas indicated by the DTCs. Task Completed ☐

17. To monitor the data stream, warm up the engine. Task Completed ☐

18. Turn the ignition switch ON. Task Completed ☐

19. Turn ON the scan tool. Task Completed ☐

20. Work through the menu on the tool until its screen reads "DATA LIST". Task Completed ☐

21. Monitor the activity of the system and record any unusual activity. Task Completed ☐

22. Summarize the results of your testing.

____

____

____

____

Problems Encountered

____

____

____

Instructor's Comments

____

____

____

ENGINE PERFORMANCE JOB SHEET 20

Checking Common Sensors

Name ______________________ Station ______________ Date ______________

NATEF Correlation

This Job Sheet addresses the following NATEF task:

B.6. Inspect and test computerized engine control system sensors, power train control module (PCM), actuators, and circuits using a graphing multimeter (GMM)/digital storage oscilloscope (DSO); perform necessary action.

Objective

Upon completion of this job sheet, you will be able to properly inspect and test computerized engine control system sensors, power train control modules (PCM), actuators, and circuits.

Tools and Materials

Digital multimeter

Protective Clothing

Goggles or safety glasses with side shields

Describe the vehicle being worked on:

Year ________ Make ____________________ Model ____________________

VIN ____________________ Engine type and size ____________________

Describe general operating condition:

____

ECT Sensors

PROCEDURE

1. Describe the location of the ECT sensor.

____

____

____

2. What color are the wires that are connected to the sensor?

____

____

3. Record the resistance specifications for a normal ECT sensor for this vehicle.

____

____

____

____

____

____

____

____

4. Disconnect the electrical connector to the sensor. Task Completed ☐

5. Measure the resistance of the sensor. It was _________ ohms at approximately ____________ °F.

6. Conclusions from the above test:

____

____

____

____

TP Sensors

PROCEDURE

1. Describe the type of lab scope you are using:

   Model ____

2. Connect the lab scope across the TP sensor.

3. With the ignition on, move the throttle from closed to fully open and then allow it to close slowly. Task Completed ☐

4. Observe the trace on the scope while moving the throttle. Describe what the trace looked like.

____

____

____

5. Based on the waveform of the TP sensor, what can you tell about the sensor?

____

____

6. With a voltmeter, measure the reference voltage to the TP sensor. The reading should be __________ volts. It was ________ volts.

7. What is the output voltage from the sensor when the throttle is closed? _________ volts.

8. What is the output voltage of the sensor when the throttle is opened? _____ volts.

9. Move the throttle from closed to fully open and then allow it to close slowly. Describe the action of the voltmeter.

____

____

10. Conclusions from the above tests:

____

____

____

____

O_2 Sensors

PROCEDURE

1. Describe the type of Digital multimeter and lab scope you will be using:

   Models ____

2. Connect the voltmeter between the O_2 sensor wire and ground. Backprobe the connector near the O_2 sensor to connect the voltmeter to the sensor signal wire.

3. With the engine idling, record and describe the voltmeter readings:

____

____

4. What does this test tell you about the sensor?

____

____

5. Remove the sensor from the exhaust manifold. Task Completed ☐

6. Connect the voltmeter between the sensor wire and the case of the sensor. Task Completed ☐

7. Using a propane torch, heat the sensor element. Task Completed ☐

8. Observe and record the voltmeter reading:

____

____

9. Your conclusions from this test:

____

____

10. Backprobe the sensor signal wire at the computer and connect a digital voltmeter from the signal wire to ground with the engine idling.

    Record the voltmeter readings: _________________________

    ____

11. Connect the voltmeter from the sensor case to the sensor ground wire on the computer.

    Record the voltmeter readings: _________________________

    ____

12. Your conclusions from these two tests:

____

____

13. Connect the scan tool to the DLC. Task Completed ☐

14. Observe and record what happens to the voltage reading from the sensor.

____

____

15. How many cross counts were there? _________________________

16. Your explanation of the above test is:

____

____

17. Connect a lab scope to the sensor and observe the trace. Task Completed ☐

18. Observe and record what happens to the voltage reading from the sensor.

____

____

19. How many cross counts were there? _________________________

20. Your explanation of the above test is:

____

____

MAP Sensors

PROCEDURE

1. Describe the type of Digital multimeter and lab scope you will be using:

   Model ____

   If the MAP sensor produces an analog voltage signal, follow this procedure.

2. With the ignition switch on, backprobe the 5-volt reference wire. Task Completed ☐

3. Connect a voltmeter from the reference wire to ground. The reading is ______ volts.

4. If the reference wire is not supplying the specified voltage, what should be checked next? ____

5. With the ignition switch on, connect the voltmeter from the sensor ground wire to the battery ground. Task Completed ☐

6. What is the measured voltage drop? ________________ volts

7. What does this indicate?

   ____

   ____

8. Backprobe the MAP sensor signal wire and connect a voltmeter from this wire to ground with the ignition switch on. What is the measured voltage?

   ________________ volts

9. What does this indicate?

   ____

   ____

10. How do you determine the barometric pressure based on these voltage readings?

    ____

    ____

    ____

    ____

11. Turn the ignition switch on and connect a voltmeter to the MAP sensor signal wire. Task Completed ☐

12. Connect a vacuum hand pump to the MAP sensor vacuum connection and apply 5 inches of vacuum to the sensor. Record the voltage reading:

    _______ volts.

NOTE: *On some MAP sensors, the sensor voltage signal should change 0.7 to 1.0 volt for every 5 inches of vacuum change applied to the sensor. Always use the vehicle manufacturer's specifications. If the barometric pressure voltage signal was 4.5 volts with 5 inches of vacuum applied to the MAP sensor, the voltage should be 3.5 volts to 3.8 volts. When 10 inches of vacuum is applied to the sensor, the voltage signal should be 2.5 volts to 3.1 volts. Check the MAP sensor voltage at 5-inch intervals from 0 to 25 inches.* If the MAP sensor voltage is not within specifications at any vacuum, replace the sensor.

13. Record the results of all vacuum checks:

____

____

14. What did these tests indicate?

____

____

15. Connect the scope to the MAP output and a good ground. Task Completed ☐

16. Accelerate the engine and allow it to return to idle. Observe and describe the trace:

____

____

17. What did the trace show about the sensor?

____

____

____

NOTE: *If the MAP sensor produces a digital voltage signal of varying frequency, check the 5-volt reference wire and the ground wire with the same procedure used on other MAP sensors. This sensor diagnosis is based on the use of a MAP sensor tester that changes the MAP sensor varying frequency voltage to an analog voltage.*

Follow these steps to test the MAP sensor voltage signal:

1. Turn off the ignition switch, and disconnect the wiring connector from the MAP sensor. Task Completed ☐
2. Connect the connector on the MAP sensor tester to the MAP sensor. Task Completed ☐
3. Connect the MAP sensor tester battery leads to a 12-volt battery. Task Completed ☐
4. Connect a pair of digital voltmeter leads to the MAP tester signal wire and ground. Task Completed ☐
5. Turn on the ignition switch and observe the barometric pressure voltage signal on the meter. Observe and record the voltmeter readings:

____

____

____

6. Supply the specified vacuum to the MAP sensor with a hand vacuum pump. Task Completed ☐
7. Observe the voltmeter reading at each specified vacuum. Record the readings:

____

____

____

8. What do these readings indicate?

____

____

____

Problems Encountered

____

____

____

Instructor's Comments

____

____

____

ENGINE PERFORMANCE JOB SHEET 21

Electronic Service Information

Name ______________________ Station ______________ Date ______________

NATEF Correlation

This Job Sheet addresses the following NATEF task:

B.8. Access and use service information to perform step-by-step diagnosis.

Objective

Upon completion of this job sheet, you will be able to access and use electronic service information (ESI).

Tools and Materials

Computer with ESI available

Protective Clothing

Goggles or safety glasses with side shields

Describe the vehicle being worked on:

Year ________ Make ____________________ Model ____________________

VIN ____________________ Engine type and size ____________________

PROCEDURE

1. Describe the computer and operating system you are using.

   ____

   ____

2. What is the source of the information you will be retrieving?

   ____

   ____

3. Enter into the software package the vehicle information requested. Task Completed ☐

4. What area of the vehicle do you want to retrieve information about, or what area was assigned to you?

   ____

   ____

   ____

5. Select that area on the computer. Task Completed ☐

6. View the table of contents for that area and select the category you are researching. What is the category?

____

____

7. Select the section of that category you are researching. What is the section?

____

____

8. Select the type of information you want about this section. What do you want (TSBs, procedures, tips, etc.)?

____

____

9. Describe the procedure given in the service manual for discovering the cause of a problem with the input from a throttle position sensor (normally code P0120).

____

____

Problems Encountered

____

____

____

Instructor's Comments

____

____

____

ENGINE PERFORMANCE JOB SHEET 22

Diagnosis of Power Supply and Ground Circuits

Name ______________________ Station ______________ Date ______________

NATEF Correlation

This Job Sheet relates to the following NATEF task:

B.9. Diagnose driveability and emissions problems resulting from malfunctions of interrelated systems (cruise control, security alarms, suspension controls, traction controls, A/C, automatic transmissions, non-OEM-installed accessories, and similar systems); determine necessary action.

Objective

Upon completion of this job sheet, you will be able to inspect and test power and ground circuits and connections.

Tools and Materials

Service manual
Wiring diagram for the vehicle
Lab scope
Digital multimeter

Protective Clothing

Goggles or safety glasses with side shields

Describe the vehicle being worked on:

Year ________ Make __________________ Model ____________________

VIN ____________________ Engine type and size ____________________

PROCEDURE

1. All PCMs (OBD II and earlier designs) cannot operate properly unless they have good ground connections and the correct voltage at the required terminals. A wiring diagram for the vehicle being tested must be used for these tests. Backprobe the battery terminal at the PCM with the ignition switch off and connect a digital voltmeter from this terminal to ground. How many volts did you measure?

   ____

2. The voltage at this terminal should be 12 volts. If 12 volts are not available at this terminal, check the computer fuse and related circuit. Is the fuse good?

   ____

3. Turn on the ignition switch and connect the red voltmeter lead to the other battery terminals at the PCM with the black lead still grounded. The voltage measured at these terminals should also be 12 volts with the ignition switch on. How many volts did you measure?

____

4. If the specified voltage is not available, test the voltage supply wires to these terminals. These terminals may be connected through fuses, fuse links, or relays. If you needed to check these, what were the results?

____

____

5. Computer ground wires usually extend from the computer to a ground connection on the engine or battery. With the ignition switch on, connect a digital voltmeter from the battery ground to the computer ground. The voltage drop across the ground wires should be 30 millivolts or less. If the voltage reading is greater than that or more than that specified by the manufacturer, repair the ground wires or connection. Summarize the results of this check.

____

____

6. Check the negative and positive terminals and cables of the battery. Conduct a voltage drop test across each. Summarize the results of this check.

____

____

7. Conduct a voltage drop test for the grounds at five of the system's sensors and actuators. List the five components whose ground you checked and summarize the results of each check.

____

____

____

____

____

8. Check the ground of the same system components with a lab scope. Poor grounds can allow EMI or noise to be present on the reference voltage signal. This noise causes small changes in the voltage going to the sensor. Therefore, the output signal from the sensor will also have these voltage changes. The computer will try to respond to these small rapid changes, which can cause a driveability problem. Summarize the results of these checks.

9. The traces on the scope should have been flat. If noise is present, move the scope's negative probe to a known good ground. If the noise disappears, the sensor's ground circuit is bad or has resistance. If the noise is still present, the voltage feed circuit is bad or there is EMI in the circuit from another source, such as the AC generator. Summarize the results of these checks.

10. If a voltage trace has a large spike and the circuit is fitted with a resistor to limit noise, the resistor may be bad. Clamping diodes are used on devices like A/C compressor clutches to eliminate voltage spikes. If the diode is bad, a negative spike will result. Capacitors or chokes are used to control noise from a motor or generator. Test the suspected component and summarize the results of that check.

Problems Encountered

Instructor's Comments

ENGINE PERFORMANCE JOB SHEET 23

Handling Static Sensitive Devices

Name ______________________ Station ______________ Date ______________

NATEF Correlation

This Job Sheet relates to the following NATEF task:

B.9. Diagnose driveability and emissions problems resulting from malfunctions of interrelated systems (cruise control, security alarms, suspension controls, traction controls, A/C, automatic transmissions, non-OEM-installed accessories, and similar systems); determine necessary action.

Objective

Upon completion of this job sheet, you will be able to practice recommended precautions when handling static sensitive devices.

Tools and Materials

Service manual

Protective Clothing

Goggles or safety glasses with side shields

Describe the vehicle being worked on:

Year ________ Make ____________________ Model ____________________

VIN ____________________ Engine type and size ____________________

PROCEDURE

1. Refer to the service manual to identify the coding or labeling used by the manufacturer to warn technicians that some components are sensitive to electrostatic discharge. Describe the method of notification used on this vehicle.

   ____

   ____

   ____

2. When handling any electronic part, especially those that are static sensitive, always touch a known good ground before handling the part. This should be repeated while handling the part and more frequently after sliding across a seat, sitting down from a standing position, or walking a distance.

   Task Completed ☐

3. Avoid touching the electrical terminals of the part, unless you are instructed to do so in the written service procedures. It is good practice to keep your fingers off all electrical terminals, because the oil from your skin can cause corrosion. Task Completed ☐

4. When you are using a voltmeter, always connect the negative meter lead first. Task Completed ☐

5. Do not remove a part from its protective package until it is time to install the part. Task Completed ☐

6. Before removing the part from its package, ground yourself and the package to a known good ground on the vehicle. Task Completed ☐

Problems Encountered

____

____

____

Instructor's Comments

____

____

____

ENGINE PERFORMANCE JOB SHEET 24

Diagnosing Related Systems

Name ______________________ Station ______________ Date ______________

NATEF Correlation

This Job Sheet addresses the following NATEF task:

B.9. Diagnose driveability and emissions problems resulting from malfunctions of interrelated systems (cruise control, security alarms, suspension controls, traction controls, A/C, automatic transmissions, non-OEM-installed accessories, and similar systems); determine necessary action.

Objective

Upon completion of this job sheet, you will be able to diagnose driveability and emissions problems resulting from failures of interrelated systems.

Tools and Materials

Service manual

Protective Clothing

Goggles or safety glasses with side shields

Describe the vehicle being worked on:

Year __________ Make ______________________ Model ______________________

VIN ______________________ Engine type and size ______________________

PROCEDURE

NOTE: *Whenever there is a customer complaint that relates to how the car operates, always check the systems that could cause the problem before diving into the engine and its systems. This job sheet is divided into the systems that could cause driveability problems. There is no attempt made to cover all of the possible problems that will affect the way a vehicle operates. Rather, what is covered are the basic systems and components that are common causes of driveability problems.*

Clutches

1. A bucking or jerking sensation can be felt if the clutch slips or its splines to the transmission shaft are worn.
2. If the clutch does not release, the engine may be hard to start or perhaps will not crank at all unless the transmission is in neutral.

3. To check for clutch drag, start the engine, depress the clutch pedal completely, and shift the transmission into first gear. Do not release the clutch. Then, shift the transmission into neutral and wait 5 seconds before attempting to shift smoothly into reverse. It should take no more than 5 seconds for the clutch disc, input shaft, and transmission gears to come to a complete stop after disengagement. This period is normal and should not be mistaken for clutch drag. Clutch drag would be made evident by gear clashing when the transmission was shifted into reverse.
4. An unbalanced clutch assembly may cause the engine to run rough. This same problem can be caused by a loose flywheel or pressure plate.
5. Slipping can be caused by a contaminated clutch disc or incorrect clutch pedal freeplay.
6. To verify that there is slippage, set the parking brake and disengage the clutch. Shift the transmission into third gear, and increase the engine speed to about 2000 rpm. Slowly release the clutch pedal until the clutch engages. The engine should stall immediately. If it does not stall within a few seconds, the clutch is slipping.
7. If any of these seem to be a likely cause of the engine performance problem, diagnose this system and summarize the results here.

____

____

Torque Converters

1. A bad torque converter may provide less torque multiplication and lead to a customer's complaint of a lack of power.
2. An unbalanced or loose torque converter and/or flex plate will cause a vibration that will make the engine seem to run rough or have a misfire.
3. If the converter clutch doesn't lock during cruising speeds, the customer will notice a decrease in fuel economy. Check the lockup circuit and the converter control circuit to determine why the clutch is not locking.
4. If the clutch locks too soon or remains locked throughout all engine speeds, there may be a lack of power and poor acceleration. If the clutch is locked while the engine is idling, the engine will stall. If this situation exists, the engine may also stall during deceleration.
5. If the converter locks too soon, there may be a surge or bucking as the clutch locks.
6. If any of these seem to be a likely cause of the engine performance problem, diagnose this system and summarize the results here.

____

____

Automatic Transmissions

1. A transmission that shifts too early or too late will affect fuel consumption and overall performance. When the transmission shifts too early, the available power to the wheels decreases, causing poor acceleration. When the gears change later than normal, the customer may complain about noise or may notice that fuel consumption has increased.

2. If the transmission slips, there will be a lack of power, poor acceleration, and poor fuel economy. Sometimes when there is slippage, the engine will seem to lurch or buck once the gear is finally fully engaged. There is a power loss and a sudden engagement of the gear; this causes a surge of power.

3. The engine can seem to run rough if the transmission (automatic or manual) mounts are loose or faulty. This is also true of bad engine mounts. The mounts dampen the normal vibration of an engine. If the mounts are bad, little or no dampening takes place.

4. If any of these seem to be a likely cause of the engine performance problem, diagnose this system and summarize the results here.

   ____

   ____

Driveline

1. Driveline problems can set up a vibration that may appear as a rough-running engine, especially at cruising or steady vehicle speeds.

2. A failed U-joint or damaged drive shaft can exhibit a variety of symptoms. A clunk that is heard when the transmission is shifted into gear is the most obvious. You can also encounter unusual noise, roughness, or vibration.

3. To help differentiate a potential drivetrain problem from other common sources of noise or vibration, it is important to note the speed and driving conditions at which the problem occurs. As a general guide, a worn U-joint is most noticeable during acceleration or deceleration and is less speed sensitive than an unbalanced tire (commonly occurring in the 30 to 60 mph range) or a bad wheel bearing (more noticeable at higher speeds).

4. If any of these seem to be a likely cause of the engine performance problem, diagnose this system and summarize the results here.

   ____

   ____

Heating and Air Conditioning

1. Power loss when the air conditioning is running is normal. However, too much power loss suggests the need for further diagnosis of the compressor and/or compressor drive. There may also be a change in speed as the compressor is engaged and disengaged. Nothing is wrong; this is normal because of the immediate load put on the engine.

2. If this seems to be a likely cause of the engine performance problem, diagnose this system and summarize the results here.

____

____

Speed Control Systems

1. If the cruise control unit has a hard time maintaining a fixed speed, the engine speed will surge up and down. To verify this problem as the cause of an engine surge at cruising speeds, simply allow the vehicle to cruise at highway speeds with and without the cruise control engaged. If the cruise control is at fault, the surging will only occur when it is engaged.
2. If this seems to be a likely cause of the engine performance problem, diagnose this system and summarize the results here.

____

____

Brake Systems

1. Failure of the brakes to release is often caused by a tight or misaligned connection between the power unit and the brake linkage. Broken pistons, diaphragms, bellows, or return springs can also cause this problem. To help pinpoint the problem, loosen the connection between the master cylinder and the brake booster. If the brakes release, the problem is caused by internal binding in the vacuum unit. If the brakes do not release, look for a crimped or restricted brake line or similar problem in the hydraulic system.
2. Proper adjustment of the master cylinder pushrod is necessary to ensure proper operation of the power brake system. A pushrod that is too long causes the master cylinder piston to close off the replenishing port, preventing hydraulic pressure from being released and resulting in brake drag. A pushrod that is too short causes excessive brake pedal travel and causes groaning noises to come from the booster when the brakes are applied.
3. Parking brakes can also be a cause of poor performance and bad fuel economy. If the parking brakes do not release, the rear brakes will drag. Even if the brakes are only slightly dragging, driveability will be affected.
4. If any of these seem to be a likely cause of the engine performance problem, diagnose this system and summarize the results here.

____

____

Wheels and Tires

1. Tires should be inflated to the recommended amount of pressure. The ideal amount of pressure is listed on the identification decal of the vehicle and is given in the vehicle's owner's manual. Underinflated tires have more rolling resistance and therefore will cause fuel economy to decrease, as well as available power. Tires that are only a little low will still have a negative effect on fuel economy. Check the pressure in the tires and look at the tires for outside wear. If the tires are severely worn, tell the customer that they should be replaced and explain why.

2. If replacement tires have been installed on the vehicle and they are not the same size as the original equipment, fuel economy and engine performance will be affected. The change in tire diameter changes the overall gear ratio of the drivetrain. The overall drive gear ratio is determined by the gear ratios in the drivetrain and the circumference of the tire. The circumference of the tire determines how many times it will rotate during a mile of travel.

3. If any of these seem to be a likely cause of the engine performance problem, diagnose this system and summarize the results here.

____

____

Problems Encountered

____

____

____

Instructor's Comments

____

____

____

ENGINE PERFORMANCE JOB SHEET 25

Using a Scan Tool to Test Actuators

Name ______________________ Station ______________ Date ______________

NATEF Correlation:

This Job Sheet addresses the following NATEF task:

B.10. Perform active tests of actuators using scan tool; determine necessary action.

Objective

Upon completion of this job sheet, you will be able to check the operation of various actuators with a scan tool.

Tools and Materials

Service manual

Scan tool

Vehicle with OBD II

Protective Clothing

Goggles or safety glasses with side shields

Describe the vehicle being worked on:

Year __________ Make ______________________ Model ______________________

VIN ______________________ Engine type and size ______________________

PROCEDURE

NOTE: *The actuators or computer outputs that can be tested directly with a scan tool varies with vehicle model and accessories the vehicle is equipped with. Refer to the service manual whenever conducting these tests.*

1. What type of scan tool are you using?

____

____

2. Refer to the operating manual for the scan tool and list the actuators that can be controlled directly by the tool.

____

____

____

____

3. Which of these will you check and describe the special procedures for checking that actuator.

____

____

4. Connect the scan tool to the DLC. Task Completed ☐
5. Turn the ignition switch ON. Task Completed ☐
6. Start the engine and allow it to warm up. Task Completed ☐
7. Select the actuator from the ACTIVE TEST menu on the scan tool. Task Completed ☐
8. Activate the actuator and listen or look for the specified changes. What did you observe?

____

____

9. Refer to the service manual and list the actuators that can be monitored by observing the data stream. List them here.

____

____

____

____

10. Choose one of those actuators and list it here with the specified test conditions and results.

____

____

11. Select that data on the scan tool and observe its activity under the specified conditions. Describe your observations.

____

____

Problems Encountered

____

____

____

Instructor's Comments

____

____

____

ENGINE PERFORMANCE JOB SHEET 26

No-Start Diagnosis

Name ______________________ Station ______________ Date ______________

NATEF Correlation

This Job Sheet addresses the following NATEF tasks:

C.1. Diagnose ignition system related problems such as no-starting, hard starting, engine misfire, poor driveability, spark knock, power loss, poor mileage, and emissions concerns on vehicles with electronic ignition (distributorless) systems; determine necessary action.

C.2. Diagnose ignition system related problems such as no-starting, hard starting, engine misfire, poor driveability, spark knock, power loss, poor mileage, and emissions concerns on vehicles with distributor ignition (DI) systems; determine necessary action.

Objective

Upon completion of this job sheet, you will be able to diagnose no-start problems.

Tools and Materials

Service manual
Fuel pressure gauge
Test spark plug
Digital multimeter

Protective Clothing

Goggles or safety glasses with side shields

Describe the vehicle being worked on:

Year __________ Make ____________________ Model ____________________

VIN ____________________ Engine type and size ____________________

PROCEDURE

1. The cause of a no-start condition can be located in the air, fuel, or ignition system. Begin by inspecting the air filter and the ductwork for air. Check for blockages, because an adequate amount of air must be supplied for the air-fuel mixture. Summarize your findings.

____

____

2. Check the fuel pressure at the fuel rail. A pressurized fuel supply must be delivered to the properly operating injectors. If there is no pressure, refer to the service manual to determine if there is a switch or relay that will shut off the fuel pump if a collision or a severe jarring is detected by the switch. Summarize your findings.

____

____

3. Check for voltage signals to the injectors. The injectors must receive a trigger signal to inject the fuel. Summarize your findings.

____

____

4. If the air and fuel systems seem to be in good shape, the ignition system is a likely cause of the no-start problem. Follow these procedures (choose the procedure that matches the type of ignition the vehicle has) to determine the exact cause of the problem. Often manufacturers include a detailed trouble shooting tree in their service manuals to help locate the cause of the no-start condition. What type of ignition does the vehicle have?

____

Distributor Ignition Systems

1. Connect a 12-volt test lamp from the coil tachometer (tach) terminal to ground. Task Completed ☐

2. Turn on the ignition switch. The test light normally should be "on." If the test light is "off," there is an open circuit in the coil primary winding or in the circuit from the ignition switch to the coil battery terminal. On many Chrysler and Ford systems, the test light should be "off" because the module primary circuit is closed. Since there is primary current flow, most of the voltage is dropped across the primary coil winding. This action results in voltage that is so low at the tach terminal that it does not illuminate the test light. On these systems, if the test light is illuminated, there is an open circuit in the module or in the wire between the coil and the module. Summarize your findings.

____

____

3. Crank the engine and observe the test light. If the test light flutters while the engine is cranked, the pickup coil signal and the module are okay. When the test lamp does not flutter, the pickup and/or module are bad. A pickup is tested with an ohmmeter. If the pickup coil is satisfactory, the module is defective. Before testing the pickup, check the voltage supply to the positive primary coil terminal with the ignition switch on before the diagnosis is continued. Summarize your findings.

____

____

4. If the test light flutters, connect a test spark plug to the coil secondary wire and ground the spark plug case. Task Completed ☐

5. Crank the engine and observe the spark plug. If the test spark plug fires, the ignition coil is satisfactory. If the test spark plug does not fire, the coil is probably bad. Summarize your findings.

____

____

6. Connect the test spark plug to several spark plug wires and crank the engine while observing the spark plug. If the test spark plug fired in step 5 but does not fire at some of the spark plugs, the secondary voltage and current is leaking through a defective distributor cap, rotor, or spark plug wires, or a plug wire is open. If the test spark plug fires at all the spark plugs, the ignition system is working fine. Summarize your findings.

____

____

Electronic (Distributorless) Ignition Systems

1. Perform a spark intensity check on each coil with a test spark plug. A bright, snapping spark indicates that the secondary voltage output is good. If the spark is weak or if there is no spark, check the primary circuit, including the crankshaft and camshaft sensors. A weak or no spark condition can also be caused by an open ignition cable, bad coil, or bad ignition module. Summarize your findings.

____

____

WARNING: *Do not crank or run an EI-equipped engine with a spark plug wire completely removed from a spark plug. This action may cause leakage defects in the coils or spark plug wires.*

2. If the test plug did not fire on a pair of spark plugs connected to the same coil, test the spark plug wires connected to that coil. If these wires are good, the coil is probably bad. When the test spark plug does not fire on any plug, connect a voltmeter from the input (battery) terminal on each coil pack to ground. With the ignition switch on, the voltmeter should read 12 volts. If the voltage is less than that, test the wire from the ignition switch to the coils and check the ignition switch. Summarize your findings.

____

____

3. If a crank or cam sensor fails, the engine will not start. Both of these sensor circuits can be checked with a voltmeter or DSO. If the sensors are receiving the correct amount of voltage and have good low-resistance ground circuits, their output should be a digital signal or display a pulsing voltmeter reading while the engine is cranking. If any of these conditions do not exist, the circuit needs to be repaired or the sensor needs to be replaced. Summarize your findings.

____

____

4. If the cause of the no-start condition has not yet been found, check the ignition coil with an ohmmeter. If the winding resistance readings are not within specifications, replace the coil or coil pack. If the coils are fine, check the primary circuit for proper voltages. Summarize your findings.

____

____

Problems Encountered

Instructor's Comments

ENGINE PERFORMANCE JOB SHEET 27

Visually Inspecting an Electronic Ignition System

Name ______________________ Station ______________ Date ______________

NATEF Correlation

This Job Sheet addresses the following NATEF tasks:

C.3. Inspect and test ignition primary circuit wiring and solid state components; perform necessary action.

C.8. Inspect and test ignition system pick-up sensor or triggering devices; perform necessary action.

Objective

Upon completion of this job sheet, you will be able to inspect and test the ignition system pick-up sensor or the triggering devices. You will also be able to inspect and test the ignition primary circuit wiring and its components.

Tools and Materials

Clean rag

Appropriate service manual

A vehicle with electronic ignition (EI)

Protective Clothing

Goggles or safety glasses with side shields

Describe the vehicle being worked on:

Year __________ Make ______________________ Model ______________________

VIN ______________________ Engine type and size ______________________

Ignition type ______________________

Describe general operating condition:

____

PROCEDURE

1. Are the spark plug cables pushed tightly into the coil and onto the spark plugs? ___ Yes ___ No
2. Do the secondary cables have cracks or signs of worn insulation? ___ Yes ___ No
3. Are the boots on the ends of the secondary wires cracked or brittle? ___ Yes ___ No

4. Are there any white or grayish powdery deposits on secondary cables? ___ Yes ___ No
5. Do the coils show signs of leakage in the coil towers? ___ Yes ___ No
6. Do the coil towers show any signs of burning? ___ Yes ___ No
7. Separate the coils and inspect the underside of the coil and the ignition module wires. Are the wires loose or damaged? ___ Yes ___ No
8. Are the connections of the primary ignition system wiring tight connections? ___ Yes ___ No
9. Is the control module tightly mounted to a clean surface? ___ Yes ___ No
10. Are the electrical connections to the module corroded? ___ Yes ___ No
11. Are the electrical connections to the module loose or damaged? ___ Yes ___ No
12. Record your summary of the visual inspection; include what looked good, as well as what looked bad.

____

____

____

13. Refer to the service manual. What are the resistance specifications for the camshaft position sensor?

____

14. Using an ohmmeter, measure the resistance between the terminals. What did you read and what does this indicate?

____

____

15. Refer to the service manual. What are the resistance specifications for the crankshaft position sensor?

____

____

16. Using an ohmmeter, measure the resistance between the terminals. What did you read and what does this indicate?

____

____

17. Based on these checks, what are your recommendations?

____

____

____

Problems Encountered

Instructor's Comments

ENGINE PERFORMANCE JOB SHEET 28

Individual Component Testing

Name ______________________ Station ______________ Date ______________

NATEF Correlation

This Job Sheet addresses the following NATEF tasks:

C.3. Inspect and test ignition primary circuit wiring and solid state components; perform necessary action.

C.4. Inspect, test, and service distributor.

C.5. Inspect and test ignition system secondary circuit wiring and components; perform necessary action.

C.8. Inspect and test ignition system pick-up sensor or triggering devices; perform necessary action.

Objective

Upon completion of this job sheet, you will be able to properly inspect and test a distributor and ignition control module. You will also be able to test the ignition system's secondary circuit wiring and its components.

Tools and Materials

12-volt test light

Digital multimeter

Appropriate service manual

Lab scope (DSO)

Protective Clothing

Goggles or safety glasses with side shields

Describe the vehicle being worked on:

Year ________ Make ______________________ Model ______________________

VIN ______________________ Engine type and size ______________________

Type of ignition system ______________________

Describe general operating condition:

____

PROCEDURE

Check the following components:

Ignition Switch

1. Turn the ignition key off and disconnect the wire connector at the module. Task Completed ☐
2. Disconnect the S terminal of the starter solenoid to prevent the engine from cranking when the ignition is in the run position. Task Completed ☐

3. Turn the key to the run position. Task Completed ☐
4. With the test light, probe the red wire connection to check for voltage. Was there voltage? ___ Yes ___ No
5. Check for voltage at the bat terminal of the ignition coil. Was there voltage? ___ Yes ___ No
6. Turn the key to the start position and check for voltage at the start power wire connector at the module. Was there voltage? ___ Yes ___ No
7. Check for voltage at the bat terminal of the ignition coil. Was there voltage? ___ Yes ___ No

Conclusions:

___

___

8. Turn the ignition switch to the off position. Task Completed ☐
9. Install a small straight pin into the appropriate module's input power wire. Task Completed ☐
10. Connect the digital voltmeter's positive lead to the straight pin and ground the negative lead to the distributor base. Task Completed ☐
11. Turn the ignition to the run position. Your voltage reading is ___
12. Turn the ignition to the start position. Your voltage reading is ___

Conclusions:

___

___

Primary Resistor

1. Turn the ignition off. Task Completed ☐
2. Connect an ohmmeter across the resistor. Task Completed ☐
3. Record your readings: ___ ohms
4. The specified resistance is ___ ohms

Conclusions:

___

___

Pick-Up Coil

1. Turn the ignition off. Task Completed ☐
2. Remove the distributor cap. Task Completed ☐
3. Connect the ohmmeter to the pick-up coil terminals. Task Completed ☐
4. Record your readings: ___ ohms
5. The specified resistance is ___ ohms

Conclusions:

____

____

6. Connect the ohmmeter from one of the pick-up leads to ground. Task Completed ☐

7. Record your readings: __________ ohms

8. The specified resistance is _________ ohms

Conclusions:

____

____

9. Reinstall the distributor cap. Then connect the scope leads to the pick-up coil leads. Task Completed ☐

10. Set the scope on its lowest scale. Task Completed ☐

11. Spin the distributor shaft by cranking the engine with the ignition disabled. Task Completed ☐

12. Describe the trace shown on the scope:

____

____

Conclusions:

____

____

13. Disconnect the scope. Task Completed ☐

14. Connect a voltmeter set on its low voltage scale. Task Completed ☐

15. Describe the meter's action:

____

____

Conclusions:

____

____

Hall-Effect Sensors

1. Connect a 12-volt battery across the plus (+) and minus (–) voltage (supply current) terminals of the Hall layer. Task Completed ☐

2. Connect a voltmeter across the minus (–) and signal voltage terminals. Task Completed ☐

3. Insert a steel feeler gauge or knife blade between the Hall layer and magnet. Then remove the feeler gauge. Task Completed ☐

4. Describe what happens on the voltmeter:

____

____

Conclusions:

____

____

5. Remove the 12-volt power source and prepare the engine to run. Task Completed ☐
6. Set a lab scope on its low scale primary pattern position. Task Completed ☐
7. Connect the primary positive lead to the Hall signal lead; the negative lead should connect to ground or the ground terminal at the sensor's connector. Task Completed ☐
8. Start the engine and observe the scope. Task Completed ☐
9. Record the trace on the scope:

____

____

Conclusions:

____

____

Control Module

1. Connect one lead of the ohmmeter to the ground terminal at the module and the other lead to a good engine ground. Task Completed ☐
2. Record your readings: __________ ohms
3. The specified resistance is _________ ohms

Conclusions:

____

____

Secondary Ignition Wires

1. Remove the distributor cap with the spark plug wires attached to the cap but disconnected from the spark plugs. Task Completed ☐
2. Calibrate an ohmmeter on the X1,000 scale. Task Completed ☐
3. Connect the ohmmeter leads from the end of a spark plug wire to the distributor cap terminal inside the cap to which the plug wire is connected. Task Completed ☐
4. Record your readings: __________ ohms
5. The specified resistance is _________ ohms

Conclusions:

____

____

Spark Plugs

1. Remove the engine's spark plugs. Place them on a bench arranged according to the cylinder number. Task Completed ☐
2. Carefully examine the electrodes and porcelain of each plug. Task Completed ☐
3. Describe the appearance of each plug: Task Completed ☐

____

____

Conclusions:

____

____

4. Measure the gap of each spark plug and record your findings:

____

____

____

5. What is the specified gap? __________ inches

Conclusions:

____

____

Problems Encountered

____

____

____

Instructor's Comments

____

____

____

ENGINE PERFORMANCE JOB SHEET 29

Testing an Ignition Coil

Name ____________________ Station ______________ Date ______________

NATEF Correlation

This Job Sheet addresses the following NATEF task:

C.6. Inspect and test ignition coil(s); perform necessary action.

Objective

Upon completion of this job sheet, you will be able to properly inspect and test ignition coils.

Tools and Materials

A vehicle or a separate ignition coil

Service manual for the above vehicle or ignition coil

Digital multimeter

Protective Clothing

Goggles or safety glasses with side shields

Describe the vehicle being worked on:

Year ________ Make ____________________ Model ____________________

VIN ____________________ Engine type and size ____________________

Describe general operating condition:

____

PROCEDURE

1. Describe the general appearance of the coil.

   ____

   ____

2. Locate the resistance specifications for the ignition coil in the service manual.

   The primary winding should have ______ ohms of resistance.

   The secondary winding should have ______ ohms of resistance.

3. If the coil is still in the vehicle, disconnect all leads connected to it. Task Completed ☐

4. Connect the ohmmeter from the negative (tach) side of the coil to its container or frame. Observe the reading on the meter. The reading was ________ ohms. What does this indicate?

____

____

5. Connect the ohmmeter from the center tower of the coil to its container or frame. Observe the reading on the meter. The reading was ________ ohms. What does this indicate?

____

____

____

6. Connect the ohmmeter across the primary winding of the coil. The reading was ________ ohms. Compare this to specifications. What does this reading indicate?

____

____

____

7. Connect the ohmmeter across the secondary winding of the coil. The reading was ________ ohms. Compare this to specifications. What does this reading indicate?

____

____

____

8. Based on the above tests, what is your conclusion about the coil?

____

____

____

Problems Encountered

____

____

____

Instructor's Comments

____

____

____

ENGINE PERFORMANCE JOB SHEET 30

Setting Ignition Timing

Name ______________________ Station ______________ Date ______________

NATEF Correlation

This Job Sheet addresses the following NATEF task:

C.7. Check and adjust ignition system timing and timing advance/retard (where applicable).

Objective

Upon completion of this job sheet, you will be able to check and adjust ignition system timing and timing advance/retard.

Tools and Materials

Timing light
Tachometer
Appropriate service manual

Protective Clothing

Goggles or safety glasses with side shields

Describe the vehicle being worked on:

Year ________ Make ____________________ Model ____________________

VIN ____________________ Engine type and size ____________________

Describe general operating condition:

____

PROCEDURE

1. Refer to the service manual or underhood decal and answer the following:

   Ignition Timing Specs __________

   Source of specifications:

   ____

   Conditions that must be met before checking the timing:

   ____

   ____

   ____

   What should the idle speed be? __________ rpm

   What is the idle speed? ________ rpm

   If not within specifications, correct the idle before proceeding.

2. Connect the timing light pick-up to the number one cylinder's spark plug wire. The power supply wires on the light should be connected to the battery terminals with the proper polarity. Task Completed ☐

3. Start the engine. Task Completed ☐

4. The engine must be idling at the manufacturer's recommended rpm, and all other timing procedures must be followed. Task Completed ☐

5. Aim the timing light marks at the timing indicator, and observe the timing marks. Timing found: __________ degrees

6. If the timing mark is not at the specified location, rotate the distributor until the mark is at the specified location. Describe any difficulties you had doing this:

____

____

7. Tighten the distributor holddown bolt to the specified torque. What is the specified torque? ____________

8. Connect the vacuum advance hose and any other connectors, hoses, or components that were disconnected so that the timing procedure could be done. Task Completed ☐

Problems Encountered

____

____

____

Instructor's Comments

____

____

____

ENGINE PERFORMANCE JOB SHEET 31

Diagnosing EFI Systems

Name ______________________ Station ______________ Date ______________

NATEF Correlation

This Job Sheet addresses the following NATEF task:

D.1. Diagnose hot or cold no-starting, hard starting, poor driveability, incorrect idle speed, poor idle, flooding, hesitation, surging, engine misfire, power loss, stalling, poor mileage, dieseling, and emissions problems on vehicles with injection-type fuel systems; determine necessary action.

Objective

Upon completion of this job sheet, you will be able to diagnose hot or cold no-starting, hard starting, poor driveability, incorrect idle speed, poor idle, flooding, hesitation, surging, engine misfire, power loss, stalling, poor mileage, dieseling, and emissions problems on a vehicle with an injection-type fuel system.

Tools and Materials

Service manual

Lab scope

Noid light

Digital multimeter

Protective Clothing

Goggles or safety glasses with side shields

Describe the vehicle being worked on:

Year ________ Make ____________________ Model ____________________

VIN ____________________ Engine type and size ____________________

PROCEDURE

1. Before proceeding with specific fuel injection checks and electronic control testing, be certain the battery is in good condition, fully charged, with clean terminals and connections. Also verify that the charging and starting systems are operating properly. Make sure all fuses and fusible links are intact and all wiring harnesses are properly routed with connections free of corrosion and tightly attached. Task Completed ☐

2. Make sure all vacuum lines are in sound condition, properly routed, and tightly attached. Task Completed ☐

3. Check the PCV system to make sure it is working properly. Also make sure all emission control systems are in place, hooked up, and operating properly. Task Completed ☐

4. Check the coolant level and the operation of the cooling system and fans. Task Completed ☐

5. Verify that the ignition system is functioning properly and that the base timing and idle speed are set to specifications. Task Completed ☐

6. Make sure the gasoline in the tank is of good quality and has not been substantially cut with alcohol or contaminated with water. Task Completed ☐

7. Connect a scan tool to the vehicle and check for DTCs. Interpret the code and follow the diagnostic procedures given in the service manual to identify the cause of the problem. What codes did you retrieve?

____

____

8. Remember that trouble codes only indicate the particular circuit in which a problem has been detected. They do not pinpoint individual problems or components. If a code suggests a component is faulty, that component needs to be thoroughly checked. The following procedures are divided by major components of the EFI system. If you suspect that one or more of these are faulty, move to that section of the job sheet. What components do you need to test to complete your diagnosis of the system?

____

____

Oxygen Sensor Diagnosis

1. Make sure the engine is at normal operating temperature before testing an oxygen (O_2) sensor. Task Completed ☐

2. Refer to the vehicle's wiring diagram to identify the terminals at the sensor. Most late-model engines use heated oxygen sensors (HO_2S). These sensors have four wires connected to them. Two are for the heater and the other two are for the sensor. Identify the terminals of the O_2 sensors on this vehicle.

____

3. An O_2 sensor can be checked with a voltmeter. Connect it between the O_2 sensor wire and ground. With the engine running, the sensor's voltage should be cycling from 0 and 1 volt. If the voltage is continually high, the air-fuel ratio may be rich or the sensor may be contaminated. If the O_2 sensor voltage is continually low, the air-fuel ratio may be lean, the sensor may be defective, or the wire between the sensor and the computer may have a high-resistance problem. If the O_2 sensor voltage signal remains in a midrange position, the computer may be in open loop or the sensor may be defective. Summarize the results of this check.

____

____

4. The activity of an O_2 sensor is best monitored with a lab scope. The switching of the sensor should be seen as the sensor signal goes to lean to rich to lean continuously. Summarize the results of this check.

____

____

5. The activity of the sensor can also be monitored on a scanner. By watching the scanner while the engine is running, the O_2 voltage should move to nearly 1 volt then drop back to close to zero volts. Immediately after it drops, the voltage signal should move back up. This immediate cycling is an important function of an O_2 sensor. If the response is slow, the sensor is lazy and should be replaced. With the engine at about 2,500 rpm, the O_2 sensor should cycle from high to low 10 to 40 times in 10 seconds. When testing the O_2 sensor, make sure the sensor is heated and the system is in closed loop. Summarize the results of this check.

____

____

Diagnosing Fuel Control Problems

1. A scan tool is also an excellent way to monitor fuel control. There are many factors that determine the pulse width of the injectors but it should always respond to the O_2 readings. Fuel correction for fuel injection systems is shown on a scan tool. Block Integrator represents a short-term correction to the amount of fuel delivered during closed-loop. Block Learn makes long-term corrections. Injector pulse width is adjusted according to both Block Integrator and Block Learn. Block Integrator and Block Learn are also known as short-term fuel trim and long-term fuel trim, or adaptive memory and additive fuel factor. A scale of 0 to 255 is used for Block Integrator and Block Learn and a mid-range reading of 128 is preferred. When the integrator, or block learn, numbers are considerably above 128, the PCM is continually attempting to increase fuel; therefore, the O_2 sensor voltage signal must be continually low, or lean. If the integrator, or block learn, numbers are considerably below 128, the PCM is continually decreasing fuel, which indicates that the O_2 sensor voltage must be always high, or rich. Monitor the fuel control of the system and interpret the results.

____

____

Air Induction System Checks

1. Check the air control system for cracks and deteriorated ductwork. Also make sure all the induction hose clamps are tight and properly sealed. Look for possible air leaks in the crankcase, for example, around the dipstick tube and oil filter cap. Summarize the results of this check.

____

____

2. Check the vacuum lines and components. Summarize the results of this check.

___

___

___

Airflow Sensors

1. Remove the air intake duct from the airflow sensor to gain access to the sensor flap. Check it for binding, sticking, or scraping by rotating the sensor flap (evenly and carefully) through its operating range. It should move freely, make no noise, and feel smooth. Summarize the results of this check.

___

___

___

2. On some systems, the movement of the sensor flap turns on the fuel pump. If the system is so equipped, turn the ignition on. Do not start the engine. Move the flap toward the open position. The electric fuel pump should come on as the flap is opened. If it does not, turn off the ignition, remove the sensor harness, and check for specific resistance values with an ohmmeter at each of the sensor's terminals. Summarize the results of this check.

___

___

___

3. On other models, it is possible to check the resistance values of the potentiometer by moving the air flap, but in either case a service manual is needed to identify the various terminals and to look up resistance specifications. What is the range of acceptable resistances and how do your measurements compare with the specifications?

___

___

___

4. Use a scan tool to check the operation of a mass airflow meter or manifold pressure sensor. Summarize the results of this check.

___

___

Throttle Body Diagnosis

1. Check the inside of the throttle body assembly for accumulations of dirt, carbon, or other substances. Summarize the results of this check.

___

___

2. Check for smooth movement of the throttle linkage from idle position to the wide-open position. Also check the throttle linkage and cable for wear and looseness. Summarize the results of this check.

____

____

____

3. Check the vacuum at each vacuum port on the throttle body while the engine is idling and while it is running at a higher speed. Summarize the results of this check.

____

____

____

4. Apply vacuum to the throttle opener. Then, disconnect the TP sensor connector, and test the TP sensor with an ohmmeter connected across the appropriate terminals. Loosen the two TP sensor mounting screws and rotate the TP sensor as required to obtain the specified ohmmeter readings. Then retighten the mounting screws. If the TP sensor cannot be adjusted to obtain the proper ohmmeter readings, replace the TP sensor. Summarize the results of this check.

____

____

____

5. Operate the engine until it reaches normal operating temperature, and check the idle speed on a tachometer. The idle speed should be 700 to 800 rpm. Disconnect and plug the vacuum hose from the throttle opener. Maintain 2,500 engine rpm. Release the throttle valve and observe the tachometer reading. When the throttle linkage strikes the throttle opener stem, the engine speed should be 1,300 to 1,500 rpm. Adjust the throttle opener as necessary, and reconnect the throttle opener vacuum hose. Summarize the results of this check.

____

____

____

Fuel System Checks

1. Check fuel pressure and fuel volume. Compare your measurements with the specifications and state your conclusions about the fuel system.

____

____

____

2. If the fuel pressure slowly bleeds down, there may be a leak in the fuel pressure regulator, fuel pump check valve, or the injectors themselves. Remember, hard starting is a common symptom of system leaks. Summarize the results of this check.

____

____

____

Injector Checks

1. Warm up a gas analyzer and set the meters to ambient air. Task Completed ☐

2. With the engine warmed up, but not running, remove the air duct from the airflow sensor. Then, insert the gas analyzer's probe into the intake plenum area. Look at the HC readings on the analyzer. They should be low and drop as time passes. If an injector is leaking, the HC reading will be high and will not drop. Summarize the results of this check.

____

____

3. Another cause of a rich mixture is a leaking fuel pressure regulator. After the engine has been run, disconnect the vacuum line to the fuel pressure regulator. If there is any sign of fuel inside the hose or if fuel comes out of the hose, the regulator's diaphragm is leaking. The regulator can also be tested with a hand-operated vacuum pump. Apply 5 in. Hg to the regulator. A good regulator diaphragm will hold that vacuum. Summarize the results of this check.

____

____

4. Check to see if the injectors are receiving a signal from the PCM to fire. Remove the injector's electrical connector and check for voltage at the injector using a high impedance test light or a noid light. While cranking the engine, the light should flash if the computer is cycling the injector on and off. If the light is not flashing, the computer or connecting wires are defective. Summarize the results of this check.

____

____

5. An ohmmeter can be used to test an injector. Connect the ohmmeter across the injector terminals after the wires to the injector have been disconnected. If the meter reading is infinite, the injector winding is open. If the meter shows more resistance than the specifications call for, there is high resistance in the winding. A reading that is lower than the specifications indicates that the winding is shorted. Compare your readings with the specifications and explain your conclusions about the injectors.

____

____

6. Connect the lab scope's positive lead to the injector supply wire and the scope's negative lead to an engine ground. Set the scope to read 12 volts, then set the sweep and trigger to allow you to clearly see the "on" signal on the left and the "off" signal on the right. Make sure the entire waveform is clearly seen. Also remember that the setting may need to be changed as engine speed increases or decreases. Observe and describe the pattern of the fuel injectors.

____

____

7. When checking the injectors with a lab scope, make sure the injectors are firing at the correct time. To do this, use a dual trace scope and monitor the ignition reference signal and a fuel injector signal at the same time. The two signals should have some sort of rhythm between them. This rhythm is dependent upon several things; however, it doesn't matter what the rhythm is, it only matters that the rhythm is constant. If the injector's waveform is fine but the rhythm varies, the ignition reference sensor circuit is faulty and is not allowing the injector to fire at the correct time. Summarize the results of this check.

____

____

Problems Encountered

____

____

____

Instructor's Comments

____

____

____

ENGINE PERFORMANCE JOB SHEET 32

Inspecting the Fuel System

Name ______________________ Station ______________ Date ______________

NATEF Correlation

This Job Sheet relates to the following NATEF task:

D.1. Diagnose hot or cold no-starting, hard starting, poor driveability, incorrect idle speed, poor idle, flooding, hesitation, surging, engine misfire, power loss, stalling, poor mileage, dieseling, and emissions problems on vehicles with injection-type fuel systems; determine necessary action.

Objective

Upon completion of this job sheet, you will be able to inspect the fuel tank and fuel cap, fuel lines, fittings, and hoses of a vehicle.

Tools and Materials

None

Protective Clothing

Goggles or safety glasses with side shields

Describe the vehicle being worked on:

Year ________ Make ____________________ Model ____________________

VIN ____________________ Engine type and size ____________________

PROCEDURE

1. The fuel tank should be inspected for leaks; road damage; corrosion and rust on metal tanks; loose, damaged, or defective seams; loose mounting bolts; and damaged mounting straps. Summarize your findings.

   ____

   ____

2. Fuel lines should be inspected for holes, cracks, leaks, kinks, or dents. Summarize your findings.

   ____

   ____

3. Steel tubing should be inspected for leaks, kinks, and deformation. This tubing should also be checked for loose connections and proper clamping to the chassis. If the fuel tubing threaded connections are loose, they must be tightened to the specified torque. Some threaded fuel line fittings contain an O-ring. If the fitting is removed, the O-ring should be replaced. Summarize your findings.

____

____

4. Nylon fuel pipes should be inspected for leaks, nicks, scratches and cuts, kinks, melting, and loose fittings. If these fuel pipes are kinked or damaged in any way, they must be replaced. Nylon fuel pipes must be secured to the chassis at regular intervals to prevent fuel pipe wear and vibration. Summarize your findings.

____

____

5. Rubber fuel hose should be inspected for leaks, cracks, cuts, kinks, oil soaking, and soft spots or deterioration. If any of these conditions are found, the fuel hose should be replaced. Summarize your findings.

____

____

Problems Encountered

____

____

____

Instructor's Comments

____

____

____

ENGINE PERFORMANCE JOB SHEET 33

Checking Fuel for Contaminants

Name ______________________ Station ______________ Date ______________

NATEF Correlation

This Job Sheet addresses the following NATEF task:

D.2. Check fuel for contaminants and quality; determine necessary action.

Objective

Upon completion of this job sheet, you will be able to check the fuel on a vehicle for contaminants and quality.

Tools and Materials

Calibrated cylinder

Protective Clothing

Goggles or safety glasses with side shields

PROCEDURE

1. Pump gasoline may contain a small amount of alcohol, normally up to 10%. However, if the amount is greater than that, problems may result such as fuel system corrosion, fuel filter plugging, deterioration of rubber fuel system components, and a lean air-fuel ratio. These fuel system problems caused by excessive alcohol in the fuel may cause driveability complaints such as lack of power, acceleration stumbles, engine stalling, and no-start. If the correct amount of fuel is being delivered to the engine and there is evidence of a lean mixture, check for vacuum leaks in the intake, then check the gasoline's alcohol content. Did you find any vacuum leaks in the system? Where?

   ____

   ____

2. Obtain a 100-milliliter (mL) cylinder graduated in 1-mL divisions. Task Completed ☐
3. Fill the cylinder to the 90-mL mark with gasoline. Task Completed ☐
4. Add 10 mL of water to the cylinder so it is filled to the 100-mL mark. Task Completed ☐
5. Install a stopper in the cylinder, and shake it vigorously for 10 to 15 seconds. Task Completed ☐
6. Carefully loosen the stopper to relieve any pressure. Task Completed ☐
7. Install the stopper and shake vigorously for another 10 to 15 seconds. Task Completed ☐
8. Carefully loosen the stopper to relieve any pressure. Task Completed ☐

9. Place the cylinder on a level surface for 5 minutes to allow liquid separation. Task Completed ☐

10. Any alcohol in the fuel is absorbed by the water and settles to the bottom. If the water content in the bottom of the cylinder exceeds 10 mL, there is alcohol in the fuel. Task Completed ☐

11. Were there any particles that settled to the bottom? Describe them.

____

____

____

12. What are your service recommendations?

____

____

____

Problems Encountered

____

____

____

Instructor's Comments

____

____

____

ENGINE PERFORMANCE JOB SHEET 34

Checking Fuel Pumps

Name ______________________ Station ______________ Date ______________

NATEF Correlation

This Job Sheet addresses the following NATEF task:

D.3. Inspect and test fuel pumps and pump control systems for pressure, regulation, and volume; perform necessary action.

Objective

Upon completion of this job sheet, you will be able to inspect and test mechanical and electrical fuel pumps.

Tools and Materials

Clean shop rags
Approved gasoline container
Pressure gauge with adapters
Bleed hose
Hand-operated vacuum pump
Service manual

Protective Clothing

Goggles or safety glasses with side shields

Describe the vehicle being worked on:

Year ________ Make __________________ Model ____________________

VIN ____________________ Engine type and size ____________________

PROCEDURE

NOTE: *Since electronic fuel injection systems have a residual fuel pressure, this pressure must be relieved before disconnecting any fuel system component. Failure to relieve the fuel pressure on electronic fuel injection (EFI) systems prior to fuel system service may result in gasoline spills, serious personal injury, and expensive property damage.*

1. Disconnect the negative battery cable to avoid fuel discharge if an accidental attempt is made to start the engine. Task Completed ☐
2. Loosen the fuel tank filler cap to relief any fuel tank vapor pressure. Task Completed ☐
3. Wrap a shop towel around the fuel pressure test port on the fuel rail and remove the dust cap from this valve. Task Completed ☐
4. Connect the fuel pressure gauge to the fuel pressure test port on the fuel rail. Task Completed ☐

5. Install the bleed hose on the gauge in an approved gasoline container and open the gauge bleed valve to relieve fuel pressure from the system into the gasoline container. Be sure all the fuel in the bleed hose is drained into the gasoline container. Task Completed ☐

6. Describe any problems encountered while following this procedure:

____

____

On EFI systems that do not have a fuel pressure test port, such as most throttle body injection (TBI) systems, follow these steps for fuel system pressure relief:

1. Loosen the fuel tank filler cap to relieve any tank vapor pressure. Task Completed ☐
2. Remove the fuel pump fuse. Task Completed ☐
3. Start and run the engine until the fuel is used up in the fuel system and the engine stops. Task Completed ☐
4. Engage the starter for 3 seconds to relieve any remaining fuel pressure. Task Completed ☐
5. Disconnect the negative battery terminal to avoid possible fuel discharge if an accidental attempt is made to start the engine. Task Completed ☐
6. Describe any problems encountered while following this procedure:

____

____

To check pressure on an electric fuel pump:

1. Look up the specifications for the fuel pump. Fuel pump pressure specifications: ____________________ psi
2. Carefully inspect the fuel rail and injectors for signs of leaks. Record findings:

____

____

3. Connect the fuel pressure tester to the Schrader valve on the fuel rail. Task Completed ☐
4. Connect a hand-operated vacuum pump to the fuel pressure regulator. Task Completed ☐
5. Turn the ignition on and observe the fuel pressure readings. Your readings are: ___________ psi.
6. Compare the readings to specifications. What is indicated by the readings?

____

____

7. Create a vacuum at the pressure regulator with the vacuum pump. Task Completed ☐

8. What happened to the fuel pressure?

____

____

9. Your service recommendations.

____

____

To check pressure on a mechanical fuel pump:

1. Check the fuel pump for gasoline leaks. If gasoline is leaking at the line fittings, these fittings should be tightened to the specified torque. If the fuel leak is still present, replace the fittings and/or fuel line. If gasoline is leaking from the vent opening in the pump housing, the pump diaphragm is leaking, and fuel pump replacement is necessary. Summarize your findings.

____

____

2. If engine oil is leaking from the vent opening in the pump housing and the PCV system is working properly, the push rod seal is worn, and pump replacement is required. A loose pivot pin may cause oil leaks between the pin and the pump housing. When oil is leaking between the pump housing and the engine block, the pump mounting bolts should be tightened to the specified torque. If the oil leak continues, the gasket between the pump housing and the engine block must be replaced. Summarize your findings.

____

____

3. The pump should be checked for excessive noise. If it makes a clicking noise when the engine is operating at a fast idle, the small spring between the rocker arm and the pump housing is broken or weak. Summarize your findings.

____

____

4. If engine performance indicates inadequate fuel, the pump should be tested for pressure and volume. Normally the procedure for checking a mechanical fuel pump begins by connecting the fuel pump tester between the carburetor inlet fuel line and the inlet nut. Compare the recorded fuel pump pressure and the fuel volume against the specifications for the vehicle. If the pressure or volume is less than specified, check the fuel filter for restrictions, and check the fuel lines for restrictions and leaks before replacing the fuel pump. Summarize your findings.

____

____

Problems Encountered

Instructor's Comments

ENGINE PERFORMANCE JOB SHEET 35

Checking EFI Fuel Delivery Circuit

Name ______________________ Station ______________ Date ______________

NATEF Correlation

This Job Sheet addresses the following NATEF task:

D.3. Inspect and test fuel pumps and pump control systems for pressure, regulation and volume; perform necessary action.

Objective

Upon completion of this job sheet, you will be able to inspect and test fuel pressure regulation systems and components of injection-type fuel systems.

Tools and Materials

Clean shop rags

Pressure gauge with adapters

Approved gasoline container

Hand-operated vacuum pump

Protective Clothing

Goggles or safety glasses with side shields

Describe the vehicle being worked on:

Year __________ Make ____________________ Model ____________________

VIN ____________________ Engine type and size ____________________

Describe general operating condition:

____

PROCEDURE

1. Look up the specifications for the fuel pump. Fuel pump pressure specifications: ______________ psi
2. Carefully inspect the fuel rail and injectors for signs of leaks. Record findings:

   ____

   ____
3. Connect the fuel pressure tester to the Schrader valve on the fuel rail. Task Completed ☐
4. Connect a hand-operated vacuum pump to the fuel pressure regulator. Task Completed ☐

5. Turn the ignition on and observe the fuel pressure readings. Your readings are: ________________ psi.
6. Compare the readings to specifications. What is indicated by the readings?

____

____

7. Create a vacuum at the pressure regulator with the vacuum pump. Task Completed ☐
8. What happened to the fuel pressure?

____

____

9. Conclusions:

____

____

Problems Encountered

____

____

____

Instructor's Comments

____

____

____

ENGINE PERFORMANCE JOB SHEET 36

Removing a Fuel Filter on an EFI Vehicle

Name ______________________ Station ______________ Date ______________

NATEF Correlation

This Job Sheet addresses the following NATEF task:

D.4. Replace fuel filters.

Objective

Upon completion of this job sheet, you will be able replace a fuel filter.

Tools and Materials

Basic hand tools
Shop towels
Fuel pressure gauge set
Approved gasoline container

Protective Clothing

Goggles or safety glasses with side shields

Describe the vehicle being worked on:

Year ________ Make __________________ Model __________________

VIN __________________ Engine type and size __________________

PROCEDURE

1. Disconnect the negative cable at the battery. Task Completed ☐
2. Loosen the fuel tank filler cap to relieve any fuel tank vapor pressure. Task Completed ☐
3. Wrap a shop towel around the Schrader valve on the fuel rail and remove the dust cap from the valve. Task Completed ☐
4. Connect the fuel pressure gauge to the Schrader valve. Task Completed ☐
5. Install the free end of the gauge bleed hose into an approved gasoline container and slowly open the gauge bleed valve to relieve the fuel pressure. Task Completed ☐
6. Place the vehicle on the hoist and position the lift arms according to the manufacturer's recommendations. Then raise the vehicle. Task Completed ☐
7. Locate the fuel filter. Describe its location:

____

____

____

8. Flush the fuel filter line connectors with water and use compressed air to blow debris off and away from the connectors. Task Completed ☐

9. Follow the recommended procedures for disconnecting the fuel inlet connector. What is the procedure and what special tools are required?

____

____

____

____

10. Follow the recommended procedures for disconnecting the fuel outlet connector. What is the procedure and what special tools are required?

____

____

____

____

11. Then remove the fuel filter. Task Completed ☐

Problems Encountered

____

____

____

Instructor's Comments

____

____

____

ENGINE PERFORMANCE JOB SHEET 37

Performing a Scan Tester Diagnosis of an Idle Air Control Motor

Name ______________________ Station ______________ Date ______________

NATEF Correlation

This Job Sheet addresses the following NATEF task:

D.5. Inspect and test cold enrichment system and components; perform necessary action.

Objective

Upon completion of this job sheet, you will be able to inspect and test cold enrichment systems and components.

Tools and Materials

Basic hand tools

Scan tool

Protective Clothing

Goggles or safety glasses with side shields

Describe the vehicle being worked on:

Year ________ Make ____________________ Model ____________________

VIN ____________________ Engine type and size ____________________

PROCEDURE

1. Make sure the ignition switch is turned off. Task Completed ☐
2. Connect the scan tool leads to the battery terminals or appropriate power source. What type of scan tool did you use?

   ____

3. Turn the ignition ON, but do not start the engine. Task Completed ☐
4. Program the scan tool as required for the vehicle being tested. What information did you need to put into the scanner?

   ____

   ____

5. Use the proper adapter to connect the scan tool to the vehicle's DLC. Task Completed ☐
6. Select idle air control (IAC) motor test on the scan tool. Task Completed ☐

7. Press the appropriate scan tool button to increase engine speed and observe the engine's rpm on the scan tool. What rpm did the engine increase to?

____

8. The tool will automatically perform a non-running actuation test of the IAC hardware and software. At the low range, the IAC motor has a target range of 16 steps. The scan tool steps the motor through and monitors these steps. What were the results?

____

____

9. At the high range, the scan tool steps the motor through and monitors 112 steps. What were the results?

____

____

Problems Encountered

____

____

____

Instructor's Comments

____

____

____

ENGINE PERFORMANCE JOB SHEET 38

Servicing a Throttle Body

Name ______________________ Station ______________ Date ______________

NATEF Correlation

This Job Sheet addresses the following NATEF task:

D.6. Inspect throttle body, air induction system, intake manifold and gaskets for vacuum leaks and/or unmetered air.

Objective

Upon completion of this job sheet, you will be able to remove, service, and install throttle body.

Tools and Materials

12-volt power supply or memory keeper
Throttle body cleaner
Compressed air
OSHA-approved air nozzle
Hand tools
Service manual

Protective Clothing

Goggles or safety glasses with side shields

Describe the vehicle being worked on:

Year ________ Make ________________ Model ________________

VIN ________________ Engine type and size ________________

PROCEDURE

NOTE: *Whenever it is necessary to remove the throttle body assembly for replacement or cleaning, make sure you follow the procedures outlined by the manufacturer.*

1. Connect a 12-volt power supply or memory keeper to the cigarette lighter socket and disconnect the negative battery cable. Task Completed ☐
2. If the vehicle is equipped with an air bag, wait at least one minute. Task Completed ☐
3. Then, unbolt and remove the throttle body. Task Completed ☐
4. Once the assembly has been removed, remove all nonmetallic parts such as the TP sensor, IAC valve, throttle opener, and the throttle body gasket from the throttle body. Task Completed ☐
5. It is now safe to clean the throttle body assembly in the recommended cleaner and blow dry with compressed air. Make sure you blow out all passages in the throttle body assembly. Task Completed ☐
6. Before reinstalling the throttle body assembly, check to make sure all metal mating surfaces are clean and free from metal burrs and scratches. Task Completed ☐

7. With new gaskets and seals, install the assembly. Task Completed ☐
8. After everything that was disconnected is reconnected, reconnect the negative battery cable and disconnect the 12-volt power supply. Task Completed ☐
9. After all is reconnected and installed, adjust the throttle linkage according to the manufacturer's recommendations. Task Completed ☐

Problems Encountered

___

___

___

Instructor's Comments

___

___

___

ENGINE PERFORMANCE JOB SHEET 39

Inspect, Clean, and Test Fuel Injectors

Name ______________________ Station ______________ Date ______________

NATEF Correlation

This Job Sheet addresses the following NATEF task:

D.7. Inspect and test fuel injectors.

Objective

Upon completion of this job sheet, you will be able to properly inspect, test, and clean fuel injectors.

Tools and Materials

Fuel pressure gauge
Test spark plug
Digital multimeter
Lab scope
Noid light
Service manual

Protective Clothing

Goggles or safety glasses with side shields

Describe the vehicle being worked on:

Year ________ Make ________________ Model ________________

VIN ________________ Engine type and size ________________

PROCEDURE

1. Connect a scan tool to the vehicle and check for DTCs. Interpret the code and follow the diagnostic procedures given in the service manual to identify the cause of the problem. What codes did you retrieve?

____

____

2. Check the fuel pressure at the fuel rail. A pressurized fuel supply must be delivered to the properly operating injectors. If there is no pressure, refer to the service manual to determine if there is a switch or relay that will shut off the fuel pump if a collision or large jar is detected by the switch. Summarize your findings.

____

____

3. Check for voltage signals to the injectors; the injectors must receive a trigger signal to inject the fuel. Summarize your findings.

____

____

4. A scan tool is also an excellent way to monitor fuel control. There are many factors that determine the pulse width of the injectors but it should always respond to the O_2 readings. Fuel correction for fuel injection systems is shown on a scan tool. Block Integrator represents a short-term correction to the amount of fuel delivered during closed-loop. Block Learn makes long-term corrections. Injector pulse width is adjusted according to both Block Integrator and Block Learn. Block Integrator and Block Learn are also known as short-term fuel trim and long-term fuel trim, or adaptive memory and additive fuel factor. A scale of 0 to 255 is used for Block Integrator and Block Learn and a mid-range reading of 128 is preferred. When the integrator, or block learn, numbers are considerably above 128, the PCM is continually attempting to increase fuel; therefore, the O_2 sensor voltage signal must be continually low, or lean. If the integrator, or block learn, numbers are considerably below 128, the PCM is continually decreasing fuel, which indicates that the O_2 sensor voltage must be always high, or rich. Monitor the fuel control of the system and interpret the results.

____

____

5. Check the inside of the throttle body assembly for accumulations of dirt, carbon, or other substances. Summarize the results of this check.

____

____

6. Check for smooth movement of the throttle linkage from idle position to the wide-open position. Also check the throttle linkage and cable for wear and looseness. Summarize the results of this check.

____

____

7. Warm-up a gas analyzer and set the meters to ambient air. Task Completed ☐

8. With the engine warmed up, but not running, remove the air duct from the airflow sensor. Then, insert the gas analyzer's probe into the intake plenum area. Look at the HC readings on the analyzer. They should be low and drop as time passes. If an injector is leaking, the HC reading will be high and will not drop. Summarize the results of this check.

____

____

9. Another cause of a rich mixture is a leaking fuel pressure regulator. After the engine has been run, disconnect the vacuum line to the fuel pressure regulator. If there is a sign of fuel inside the hose or if fuel comes out of the hose, the regulator's diaphragm is leaking. The regulator can also be tested with a hand-operated vacuum pump. Apply 5 in. Hg to the regulator. A good regulator diaphragm will hold that vacuum. Summarize the results of this check.

____

____

10. Check to see if the injectors are receiving a signal from the PCM to fire. Remove the injector's electrical connector and check for voltage at the injector using a high impedance test light or a noid light. While cranking the engine, the light should flash if the computer is cycling the injector on and off. If the light is not flashing, the computer or connecting wires are defective. Summarize the results of this check.

____

____

11. An ohmmeter can be used to test an injector. Connect the ohmmeter across the injector terminals after the wires to the injector have been disconnected. If the meter reading is infinite, the injector winding is open. If the meter shows more resistance than the specifications call for, there is high resistance in the winding. A reading that is lower than the specifications indicates that the winding is shorted. Compare your readings to specifications and explain your conclusions about the injectors.

____

____

12. Connect the lab scope's positive lead to the injector supply wire and the scope's negative lead to an engine ground. Set the scope to read 12 volts, then set the sweep and trigger to allow you to clearly see the "on" signal on the left and the "off" signal on the right. Make sure the entire waveform is clearly seen. Also remember that the setting may need to be changed as engine speed increases or decreases. Observe and describe the pattern of the fuel injectors.

____

____

13. When checking the injectors with a lab scope, make sure the injectors are firing at the correct time. To do this, use a dual trace scope and monitor the ignition reference signal and a fuel injector signal at the same time. The two signals should have some sort of rhythm between them. This rhythm is dependent upon several things; however, it doesn't matter what the rhythm is, it only matters that the rhythm is constant. If the injector's waveform is fine but the rhythm varies, the ignition reference sensor circuit is faulty and is not allowing the injector to fire at the correct time. Summarize the results of this check.

____

____

CAUTION: *Often an individual injector needs to be replaced. Random disassembly of the components and improper procedures can result in damage to one of the various systems located near the injectors.*

14. The injectors are normally attached directly to a fuel rail and inserted into the intake manifold or cylinder head. They must be positively sealed because high pressure fuel leaks can cause a serious safety hazard. What normally hot components are located near the fuel injectors?

____

____

15. Prior to loosening any fitting in the fuel system, the fuel pump fuse should be removed. Where is this fuse located?

____

16. As an extra precaution, disconnect the negative cable at the battery. Task Completed ☐

17. To remove an injector, the fuel rail must be able to move away from the engine. The rail holding brackets should be unbolted and the vacuum line to the pressure regulator disconnected. Task Completed ☐

18. Disconnect the wiring harness to the injectors by depressing the center of the attaching wire clip. Task Completed ☐

19. The injectors are held to the fuel rail by a clip that fits over the top of the injector. An O-ring at the top and at the bottom of the injector seals the injector. Pull up on the fuel rail assembly. The bottom of the injectors will pull out of the manifold while the tops are secured to the rail by clips. Task Completed ☐

20. Remove the clip from the top of the injector and remove the injector unit. Task Completed ☐

21. Install new O-rings onto the new injector. Be careful not to damage the seals while installing them and make sure they are in their proper locations. Task Completed ☐

22. Install the injector into the fuel rail and set the rail assembly into place. Task Completed ☐

23. Tighten the fuel rail hold-down bolts according to the manufacturer's specifications. What are the specifications?

____

24. Reconnect all parts that were disconnected. Install the fuel pump fuse and reconnect the battery. Task Completed ☐

25. Turn the ignition switch to the run position and check the entire system for leaks. Task Completed ☐

26. After a visual inspection has been completed, conduct a fuel pressure test on the system. The results of the pressure test indicate:

    ____

    ____

    ____

    ____

Problems Encountered

____

____

____

Instructor's Comments

____

____

____

ENGINE PERFORMANCE JOB SHEET 40

Visually Inspect an EFI System

Name ______________________ Station ______________ Date ______________

NATEF Correlation

This Job Sheet addresses the following NATEF task:

D.6. Inspect throttle body, air induction system, intake manifold and gaskets for vacuum leaks and/or unmetered air.

Objective

Upon completion of this job sheet, you will be able to inspect the throttle body mounting plates, air induction and filtration systems, intake manifolds, and gaskets.

Tools and Materials

Service manual

Protective Clothing

Goggles or safety glasses with side shields

Describe the vehicle being worked on:

Year ________ Make ____________________ Model ____________________

VIN ____________________ Engine type and size ____________________

Describe general operating condition:

____

PROCEDURE

1. Is the battery in good condition, fully charged, with clean terminals and connections? ___ Yes ___ No
2. Do the charging and starting systems operate properly? ___ Yes ___ No
3. Are all fuses and fusible links intact? ___ Yes ___ No
4. Are all wiring harnesses properly routed, with connections free of corrosion and tightly attached? ___ Yes ___ No
5. Are all vacuum lines in sound condition, properly routed, and tightly attached? ___ Yes ___ No
6. Is the PCV system working properly and maintaining a sealed crankcase? ___ Yes ___ No

7. Are all emission control systems in place, hooked up and operating properly? ___ Yes ___ No
8. Is the level and condition of the coolant/antifreeze good and is the thermostat opening at the proper temperature? ___ Yes ___ No
9. Are the secondary spark delivery components in good shape, with no signs of crossfiring, carbon tracking, corrosion, or wear? ___ Yes ___ No
10. Is the base timing and idle speed set to specifications? ___ Yes ___ No
11. Is the engine in good mechanical condition? ___ Yes ___ No
12. Is the gasoline in the tank of good quality and not been substantially cut with alcohol or contaminated with water? ___ Yes ___ No
13. Does the air intake duct work have cracks or tears? ___ Yes ___ No
14. Are all of the induction hose clamps tight and properly sealed? ___ Yes ___ No
15. Are there any other possible air leaks in the crankcase? ___ Yes ___ No
16. Does the airflow sensor's flap bind, stick, or scrape when rotated through its operating range? ___ Yes ___ No
17. Are the electrical connections at the mass airflow meter or manifold pressure sensor good? ___ Yes ___ No
18. Is there carbon buildup inside the throttle bore and on the throttle plate? ___ Yes ___ No
19. Is there a clicking noise at each injector while the engine is running? ___ Yes ___ No

Problems Encountered

___

___

___

Instructor's Comments

___

___

___

ENGINE PERFORMANCE JOB SHEET 41

Checking Idle Speed

Name ______________________ Station ______________ Date ______________

NATEF Correlation:

This Job Sheet addresses the following NATEF task:

D.8. Check idle speed.

Objective

Upon completion of this job sheet, you will be able to check engine idle speed on a late-model vehicle.

Tools and Materials

Service manual

Scan tool

Protective Clothing

Goggles or safety glasses with side shields

Describe the vehicle being worked on:

Year ________ Make ____________________ Model ____________________

VIN ____________________ Engine type and size ____________________

PROCEDURE

1. Before checking the idle speed, check that the MIL is not lit. If it is, diagnose the cause. What did you find?

   ____

   ____

2. Visually inspect the air cleaner assembly and PCV system for damage. Also, check to make sure the ignition timing is correct and there are no apparent ignition problems. What did you find?

   ____

   ____

3. Does the manufacturer call for an idle learn sequence? If so, briefly explain the process.

   ____

   ____

4. Disconnect the evaporative emission (EVAP) canister purge valve connector. Task Completed ☐

5. Connect the scan tool to the date link connector (DLC). Task Completed ☐

6. Start the engine. Hold the engine at 3,000 rpm without load (in park or neutral) until the radiator fan comes on, then let it idle. Task Completed ☐

7. Check the idle speed with the headlights, blower fan, radiator fan, and air conditioner off. What was the engine speed?

____

8. Let the engine idle for 1 minute with the heater fan and air conditioner off. Task Completed ☐

9. Check the idle speed and compare to specifications. Is the idle speed correct?

____

____

10. Let the engine idle for 1 minute with the headlights, blower fan, radiator fan, and air conditioner on. Task Completed ☐

11. Check the idle speed and compare to specifications. Is the idle speed correct? What should be done to correct it?

____

____

12. Turn off the engine. Task Completed ☐

13. Reconnect the EVAP canister purge valve connector. Task Completed ☐

14. If the idle speed was not correct, check the throttle cable for binding or excessive freeplay. What did you find?

____

____

15. If the throttle cable needs adjustment, describe the procedure for doing this.

____

____

16. With the cable properly adjusted, make sure the throttle plate opens fully when you push the accelerator pedal to the floor. Also, make sure it returns to the idle position when the accelerator pedal is released. What did you find?

____

____

17. If no problems were evident, check the intake manifold and throttle body for vacuum leaks. What did you find?

____

____

Problems Encountered

____

____

____

Instructor's Comments

____

____

____

ENGINE PERFORMANCE JOB SHEET 42

Checking Fuel Delivery Circuits

Name ______________________ Station ______________ Date ______________

NATEF Correlation

This Job Sheet addresses the following NATEF task:

D.3. Inspect and test fuel pumps and pump control systems for pressure, regulation, and volume; perform necessary action.

Objective

Upon completion of this job sheet, you will be able to remove, inspect, and test the vacuum and electrical components and circuits of a fuel delivery system, as well as check the system for leaks.

Tools and Materials

Clean shop rags
Digital multimeter
Fuel pressure gauge
Service manual

Protective Clothing

Goggles or safety glasses with side shields

Describe the vehicle being worked on:

Year ________ Make ____________________ Model ____________________

VIN ____________________ Engine type and size ____________________

PROCEDURE

1. Carefully inspect the wiring and electrical connectors to the fuel pump. Record your findings.

____

____

2. Describe the outward condition of the fuel pump.

____

____

3. Carefully inspect the fuel lines and connectors for signs of leakage and damage. Record findings:

____

____

4. Look up the specifications for the fuel pump. The fuel pump pressure specifications are: ____________________ psi

5. Carefully inspect the fuel rail and injectors for signs of leaks. Record findings:

____

____

6. Connect the fuel pressure tester to the Schrader valve on the fuel rail. Task Completed ☐

7. Connect a hand-operated vacuum pump to the fuel pressure regulator. Task Completed ☐

8. Turn the ignition on and observe the fuel pressure readings. Your readings are: __________ psi.

9. Compare the readings to specifications. What is indicated by the readings?

____

____

10. Create a vacuum at the pressure regulator with the vacuum pump. Task Completed ☐

11. What happened to the fuel pressure?

____

____

12. What are your service recommendations?

____

____

Problems Encountered

____

____

____

Instructor's Comments

____

____

____

ENGINE PERFORMANCE JOB SHEET 43

Inspect Exhaust System

Name ______________________ Station ______________ Date ______________

NATEF Correlation

This Job Sheet addresses the following NATEF task:

D.9. Inspect the integrity of the exhaust manifold, exhaust pipes, muffler(s), catalytic converter(s), resonator(s), tail pipe(s), and heat shield(s); perform necessary action.

Objective

Upon completion of this job sheet, you will be able to properly inspect exhaust manifolds, exhaust pipes, mufflers, catalytic converters, resonators, tail pipes, and heat shields.

Tools and Materials

Flashlight or trouble light
Hammer or mallet
Lift
Service manual
Tachometer
Vacuum gauge

Protective Clothing

Goggles or safety glasses with side shields

Describe the vehicle being worked on:

Year ________ Make ________________ Model ________________

VIN ________________ Engine type and size ________________

Describe general operating condition:

____

PROCEDURE

1. Before doing a visual inspection, listen closely for hissing or rumbling that may indicate the beginning of exhaust system failure. With the engine idling, slowly move along the entire system and listen for leaks. Task Completed ☐

   WARNING: *Be very careful. Remember that the exhaust system gets very hot. Do not get your face too close when listening for leaks.*

   A. Did you hear any indications of a leak? ☐ Yes ☐ No

   B. If so, where? ____

   ____

   ____

2. Safely raise the vehicle. Task Completed ☐

3. With a flashlight or trouble light, check for the following: Task Completed ☐
   - Holes and road damage
   - Discoloration and rust
   - Carbon smudges
   - Bulging muffler seams
   - Interfering rattle points
   - Torn or broken hangers and clamps
   - Missing or damaged heat shields

   a. Did you detect any of these problems? ☐ Yes ☐ No

   b. If so, where? ____

   ____

   ____

4. Sound out the system by gently tapping the pipes and muffler with a hammer or mallet. A good part will have a solid metallic sound. A weak or worn-out part will have a dull sound. Listen for falling rust particles on the inside of the muffler. Mufflers usually corrode from the inside out, so the damage may not be visible from the outside. Remember that some rust spots might be only surface rust. Task Completed ☐

   a. Did you find any weak or worn-out parts? ☐ Yes ☐ No

   b. If so, which ones? ____

   ____

   ____

5. Grab the tailpipe (when it is cool) and try to move it up and down and from side to side. There should be only slight movement in any direction. If the system feels wobbly or loose, check the clamps and hangers that fasten the tailpipe to the vehicle. Task Completed ☐

   a. Did you detect any problems with the clamps or hangers? ☐ Yes ☐ No

   b. If so, where? ____

   ____

   ____

6. Check all of the pipes for kinks and dents that might restrict the flow of exhaust gases. Task Completed ☐

   a. Did you find any kinks or dents? ☐ Yes ☐ No

   b. If so, where? ____

   ____

   ____

7. Take a close look at each connection, including the one between the exhaust manifold and exhaust pipe. Task Completed ☐

   a. Did you find any white powdery deposits? ☐ Yes ☐ No

   b. If so, try tightening the bolts or replacing the gasket at that connection. Task Completed ☐

c. Check for loose connections at the muffler by pushing up on the muffler slightly. Task Completed ☐

d. If loose, try tightening them. Task Completed ☐ Not applicable ☐

8. If a visual inspection does not identify a partially restricted or blocked exhaust system, perform the following test:

a. Attach a vacuum gauge to the intake manifold. Connect a tachometer. Start the engine and observe the vacuum gauge. Task Completed ☐

It should indicate a vacuum of 16–20 in. of mercury. Does it? ☐ Yes ☐ No

b. Increase the engine's speed to 2,000 rpm and observe the vacuum gauge. Vacuum will decrease when the speed is increased rapidly, but it should stabilize at 16–21 in. of mercury and remain constant. If the vacuum does not build up to at least the idle reading, the exhaust system is restricted or blocked. Task Completed ☐

Is the system restricted or blocked? ☐ Yes ☐ No

9. Catalytic converters can overheat. Look for bluish or brownish discoloration of the outer stainless steel shell. Also, look for blistered or burned paint or undercoating above and near the converter. Task Completed ☐

Are there any signs of overheating? ☐ Yes ☐ No

10. Look up and record the part numbers of any parts discovered to be defective in previous steps. Task Completed ☐

____

____

____

____

____

____

____

Problems Encountered

____

____

____

Instructor's Comments

____

____

____

ENGINE PERFORMANCE JOB SHEET 44

Test a Catalytic Converter for Efficiency

Name ______________________ Station ______________ Date ______________

NATEF Correlation

This Job Sheet addresses the following NATEF tasks:

D.10. Perform exhaust system back-pressure test; determine necessary action.

E.3.1. Diagnose emissions and driveability problems resulting from malfunctions in the secondary air injection and catalytic converter systems; determine necessary action.

E.3.4. Inspect and test catalytic converter performance.

Objective

Upon completion of this job sheet, you will be able to properly perform exhaust system back-pressure tests, diagnose emissions and driveability problems resulting from failure of the secondary air injection and catalytic converter systems, and properly inspect and test catalytic converter systems.

Tools and Materials

Rubber mallet
Pyrometer
Pressure gauge
Propane enrichment tool
Exhaust gas analyzer
A vehicle
Hoist

Protective Clothing

Goggles or safety glasses with side shields

Describe the vehicle being worked on:

Year __________ Make ____________________ Model ____________________

VIN ____________________ Engine type and size ____________________

Describe general operating condition:

____

PROCEDURE

1. Securely raise the vehicle on a hoist. Task Completed ☐

   Make sure you have easy access to the catalytic converter and that it is not HOT.

2. Smack the exhaust pipe, by the converter, with a rubber mallet. Did it rattle?

   ____

   If it did, it needs to be replaced and there is no need to do any more testing. A rattle indicates loose substrate, which will soon rattle into small pieces.

3. If the converter passed this test, it doesn't mean it is in good shape. It should be checked for plugging or restrictions. Task Completed ☐

4. Lower the vehicle and open the hood. Task Completed ☐

5. Remove the O_2 sensor. Task Completed ☐

6. Install a pressure gauge into the sensor's bore. On some engines, it is not easy to do. On engines with a PFE, you can use its port to install your gauge. On other engines, you may need to fabricate a tester from an old O_2 sensor or air check valve. Task Completed ☐

7. After the gauge is in place, start the engine and hold the engine's speed at 2,000 rpm. Record the reading on the pressure gauge. ____________ psi.

   You are looking for exhaust pressure under 1.25 psi. A very bad restriction would give you over 2.75 psi.

8. What does your reading tell you?

   ____

   ____

   Newer cars should have pressures well under 1.25. Some older ones can be as high as 1.75 and still be good. You will notice that if you quickly rev up the engine, the pressure goes up. This is normal. Remember, do this test at 2,000 rpm, not with the throttle wide open.

9. Remove the pressure gauge, turn off the engine, allow the exhaust to cool, and then install the O_2 sensor. Task Completed ☐

10. If the converter passed this test, you can now check its efficiency. There are three ways to do this. The first way is the delta temperature test. Start the engine and allow it to warm up. With the engine running, carefully raise the vehicle using the hoist or lift. Task Completed ☐

11. With a pyrometer, measure and record the inlet temperature of the converter. The reading was __________ .

12. Now measure the temperature of the converter's outlet. The reading was __________ .

13. What was the percentage increase of the temperature at the outlet compared to the temperature of the inlet? ____________ There should be a temperature increase of about 8% or 100 degrees at the cat's outlet. If the temperature doesn't increase by 8%, replace the converter. What are your conclusions based on this test?

    ____

    ____

14. How does temperature show the efficiency of a catalytic converter?

    ____

    ____

15. Now you can do the O_2 storage test. This is based on the fact that a good converter stores oxygen. The following test is for closed-loop feedback systems. Non-feedback systems require a different procedure. Begin by disabling the air injection system. Task Completed ☐

16. Turn on your gas analyzer and allow it to warm up. Start the engine and warm the cat up as well. Task Completed ☐

17. When everything is ready, hold the engine at 2,000 rpm. Watch the exhaust readings. Record the readings:

_______HC_______CO_______CO_2___________O_2 __________NO_x

If the converter was cold, the readings should continue to drop until the converter reaches light-off temperature.

18. When the numbers stop dropping, check the oxygen levels. Check and record the oxygen level; the reading was ____________. O_2 should be about .5 to 1%. This shows the converter is using most of the available oxygen. There is one exception to this: if there is no CO left, there can be more oxygen in the exhaust. However, it still should be less than 2.5%. It is important that you get your O_2 reading as soon as the CO begins to drop. Otherwise, good converters will fail this test. The O_2 will go way over 1.25 after the CO starts to drop.

19. If there is too much oxygen left, and no CO in the exhaust, stop the test and make sure the system has control of the air-fuel mixture. If the system is in control, use your propane enrichment tool to bring the CO level up to about .5%. Now the O_2 level should drop to zero. Task Completed ☐

20. Once you have a solid oxygen reading, snap the throttle open, then let it drop back to idle. Check the oxygen. The reading was __________. It should not rise above 1.2%.

21. If the converter passes these tests, it is working properly. If the converter fails the tests, chances are that it is working poorly or not at all. The final converter test uses a principle that checks the converter as it is doing its actual job, converting CO and HC into CO_2 and water.

22. Allow the converter to warm up by running the engine. Task Completed ☐

23. Calibrate the gas analyzer and insert its probe into the exhaust pipe. If the vehicle has dual exhaust with a cross over, plug the side that the probe isn't in. If the vehicle has a true dual exhaust system, check both sides separately. Task Completed ☐

24. Turn off the engine and disable the ignition. Task Completed ☐

25. Crank the engine for 9 seconds as you pump the throttle. Look at the gas analyzer and record the CO_2 reading: __________. The CO_2 for injected cars should be over 11% and over 10% for engines with a carburetor. If you are cranking the engine and the HC goes above 1,500 ppm, stop cranking: the converter is not working. Also stop cranking once you hit your 10 or 11% CO_2 mark; the converter is good. If the converter is bad, you should see high HC and low CO_2 at the tailpipe. What are your conclusions from this test?

____

____

26. Do not repeat this test more than ONE time without running the engine in between. Task Completed ☐

27. Reconnect the ignition and start the engine. Do this as quickly as possible to cool off the converter. Task Completed ☐

Problems Encountered

____

____

____

Instructor's Comments

____

____

____

ENGINE PERFORMANCE JOB SHEET 45

Checking Boost Systems

Name ______________________ Station ______________ Date ______________

NATEF Correlation

This Job Sheet addresses the following NATEF task:

D.11. Test the operation of turbocharger/supercharger systems; determine necessary action.

Objective

Upon completion of this job sheet, you will be able to test the operation of a turbocharger/supercharger system.

Tools and Materials

Soap and water mixture

Pressure gauge

Service manual

Protective Clothing

Goggles or safety glasses with side shields

Describe the vehicle being worked on:

Year ________ Make ____________________ Model ____________________

VIN ____________________ Engine type and size ____________________

PROCEDURE

1. Check all linkages and hoses connected to the turbocharger. Record your findings.

   ____

   ____

2. Inspect the waste gate diaphragm linkage for looseness and binding, and check the hose from the waste gate diaphragm to the intake manifold for cracks, kinks, and restrictions. Record your findings.

   ____

   ____

3. Check the coolant hoses and oil line connected to the turbocharger for leaks. Record your findings.

   ____

   ____

4. When the engine is running, does the exhaust have an odor or color? Describe the exhaust.

____

____

5. Check the condition of the engine's oil. Record your findings.

____

____

6. Check all turbocharger mounting bolts for looseness. Record your findings.

____

____

7. Pay close attention to the sound of the turbocharger when it is operating. Does it make unusual noises? Record your findings.

____

____

8. Check for exhaust leaks in the turbine housing and related pipe connections. Record your findings.

____

____

9. Check the computer with a scan tool for DTCs. Record your findings.

____

____

10. Start the engine and listen to the sound the turbo system makes while you change engine speed. Record your findings.

____

____

11. Check the air cleaner and remove the ducting from the air cleaner to turbo and look for dirt buildup or damage from foreign objects. Check for loose clamps on the compressor outlet connections. Record your findings.

____

____

12. Check the engine intake system for loose bolts or leaking gaskets. Record your findings.

____

____

13. Disconnect the exhaust pipe and look for restrictions or loose material. Examine the exhaust system for cracks, loose nuts, or blown gaskets. Record your findings.

____

____

14. Rotate the turbo shaft assembly. Does it rotate freely? Are there signs of rubbing or wheel impact damage? Record your findings.

____

____

15. Visually inspect all hoses, gaskets, and tubing for proper fit, damage, and wear. Record your findings.

____

____

16. Check the low pressure, or air cleaner, side of the intake system for vacuum leaks. Record your findings.

____

____

17. On the pressure side of the system, you can check for leaks by using soapy water. After applying the soap mixture, look for bubbles to pinpoint the source of the leak. Record your findings.

____

____

18. Connect a pressure gauge to the intake manifold to check the boost pressure. Set the gauge so you can see it during a road test. Task Completed ☐

19. Road test the vehicle at the speed specified by the vehicle manufacturer and observe the boost pressure. Record your findings.

____

____

20. To check the waste gate, connect a hand pressure pump and a pressure gauge to the waste gate diaphragm. Task Completed ☐

21. Position a dial indicator against the outer end of the waste gate diaphragm rod and supply the specified pressure to the waste gate diaphragm and observe the dial indicator movement. Record your findings.

____

____

22. If the waste gate rod movement is less than specified, disconnect the rod from the waste gate valve linkage, and check the linkage for binding. Record your findings.

____

____

23. What are your service recommendations?

____

____

____

Problems Encountered

____

____

____

Instructor's Comments

____

____

____

ENGINE PERFORMANCE JOB SHEET 46

Check the Operation of a PCV System

Name ______________________ Station ______________ Date ______________

NATEF Correlation

This Job Sheet addresses the following NATEF tasks:

E.1.1. Diagnose oil leaks, emissions, and driveability problems resulting from malfunctions in the positive crankcase ventilation (PCV) system; determine necessary action.

E.1.2. Inspect, test, and service positive crankcase ventilation (PCV) filter/breather cap, valve, tubes, orifices, and hoses; perform necessary action.

Objective

Upon completion of this job sheet, you will be able to properly perform the following tasks:

1. Diagnose oil leaks, emissions, and driveability problems resulting from failure of the positive crankcase ventilation (PCV) system.
2. Inspect and test positive crankcase ventilation (PCV) filter/breather cap, valve, tubes, orifices, and hoses.

Tools and Materials

Appropriate service manual

Exhaust gas analyzer

Clean engine oil

Protective Clothing

Goggles or safety glasses with side shields

Describe the vehicle being worked on:

Year __________ Make ____________________ Model ____________________

VIN ______________________ Engine type and size ______________________

Describe general operating condition:

____

PROCEDURE

1. Describe the location of the PCV valve.

____

____

2. Run the engine until normal operating temperature is reached. Task Completed ☐

3. Record the CO reading from the exhaust analyzer. ________ %
4. Remove the PCV valve from the valve or camshaft cover. Task Completed ☐
5. Record the CO reading now. ________%
6. Explain why there was a change in CO and what service you would recommend.

____

____

____

7. Place your thumb over the end of the PCV valve. Task Completed ☐
8. Record the CO reading now. ________%
9. Explain why there was a change in CO and what service you would recommend.

____

____

____

10. Remove the valve from its hose and check for vacuum. Task Completed ☐

    Record your results.

____

____

____

11. Hold and shake the PCV valve. Task Completed ☐
12. Record your results.

____

____

13. State your conclusions about the PCV system on this engine.

____

____

____

Problems Encountered

____

____

____

Instructor's Comments

____

____

____

ENGINE PERFORMANCE JOB SHEET 47

Check the Operation of an EGR Valve

Name ______________________ Station ______________ Date ______________

NATEF Correlation

This Job Sheet addresses the following NATEF tasks:

E.2.1. Diagnose emissions and driveability problems caused by malfunctions in the exhaust gas recirculation (EGR) system; determine necessary action.

E.2.2. Inspect, test, service and replace components of the EGR system, including EGR tubing, exhaust passages, vacuum/pressure controls, filters and hoses; perform necessary action.

E.2.3. Inspect and test electrical/electronic sensors, controls, and wiring of exhaust gas recirculation (EGR) systems; perform necessary action.

Objective

Upon completion of this job sheet, you will be able to diagnose the emissions and driveability problems caused by failure of the exhaust gas recirculation (EGR) system. You will be able to inspect and test valve, valve manifold, exhaust passages, vacuum/pressure controls, filters, and hoses of exhaust gas recirculation (EGR) systems. You will also be able to inspect and test electrical/electronic sensors, controls, and wiring of exhaust gas recirculation (EGR) systems.

Tools and Materials

Hand-operated vacuum pump

Vacuum gauge

Exhaust gas analyzer

Scan tool for assigned vehicle

A vehicle with an electronic EGR transducer and solenoid

Protective Clothing

Goggles or safety glasses with side shields

Describe the vehicle being worked on:

Year ________ Make ____________________ Model ____________________

VIN ____________________ Engine type and size ____________________

Describe general operating condition:

____

PROCEDURE

1. Describe the type of EGR system found on the vehicle:

____

____

2. Inspect the hoses and connections within the EGR system. Record the results.

____

____

3. Pull the hose to the EGR transducer. Task Completed ☐
4. Run the engine and check for vacuum at the hose. Task Completed ☐
5. Turn off the engine and connect the scan tool to the DLC. Task Completed ☐
6. Insert a tee fitting into the vacuum hose and reconnect the supply line to the transducer. Connect the vacuum gauge to the tee fitting. Task Completed ☐
7. Start the engine. Task Completed ☐
8. Record the vacuum reading: __________ in. Hg
9. There should be a minimum of 15 in. Hg. Summarize your readings.

____

____

10. Actuate the solenoid with the scan tool. Task Completed ☐
11. Did the solenoid click? ___ Yes ___ No
12. Did the vacuum fluctuate with the cycling of the solenoid? ___ Yes ___ No
13. Disconnect the vacuum hose and back-pressure hose from the transducer. Task Completed ☐
14. Disconnect the electrical connector from the solenoid. Task Completed ☐
15. Plug the transducer output port. Task Completed ☐
16. Apply 1 to 2 psi of air pressure to the back-pressure port of the transducer. Task Completed ☐
17. With the vacuum pump, apply at least 12 in. Hg to the other side of the transducer. How did the transducer react?

____

____

What did this test tell you?

____

____

18. Run the engine at fast idle. Task Completed ☐

19. Insert the probe of the gas analyzer in the vehicle's tailpipe. Task Completed ☐

20. Record these exhaust levels:

_______HC_______CO_______CO_2_________O_2 _________NO_x

What was the injector pulse width at this time? _______ ms

21. Remove the vacuum hose at the EGR valve and attach the vacuum pump to the valve. Task Completed ☐

22. Apply just enough vacuum to cause the engine to run rough. Keep the vacuum at that point for a few minutes and record the readings on the gas analyzer while the engine is running rough.

_______HC_______CO_______CO_2_________O_2 _________NO_x

What was the injector pulse width at this time? _______ ms

23. Describe what happened to the emissions levels and explain why:

____

____

____

____

24. Apply the vehicle's parking brake. Task Completed ☐

25. Disconnect the vacuum pump and plug the vacuum hose to the EGR valve. Task Completed ☐

26. Firmly depress and hold the brake pedal with your left foot and place the transmission in drive. Task Completed ☐

27. Raise engine speed to about 1,800 rpm. Task Completed ☐

28. Record the readings on the gas analyzer.

_______HC_______CO_______CO_2_________O_2 _________NO_x

What was the injector pulse width at this time? _______ ms

29. Return the engine to an idle speed, place the transmission into park, then shut off the engine. Task Completed ☐

30. Describe what happened to the emissions levels and explain why:

____

____

____

____

31. Reconnect the EGR valve. Task Completed ☐

32. Apply the vehicle's parking brake. Start the engine. Task Completed ☐

33. Firmly depress and hold the brake pedal with your left foot and place the transmission in drive. Task Completed ☐

34. Raise the engine speed to about 1,800 rpm. Task Completed ☐

35. Record the readings on the gas analyzer.

_______HC_______CO_______CO_2__________O_2 __________NO_x

What was the injector pulse width at this time? _______ ms

36. Return the engine to an idle speed, place the transmission into park, then shut off the engine. Task Completed ☐

37. Describe what happened to the emissions levels and explain why:

____

____

____

____

38. Use the scan tool and retrieve any DTCs. Record any that were displayed.

____

39. Clear the codes. Task Completed ☐

Problems Encountered

____

____

____

Instructor's Comments

____

____

____

ENGINE PERFORMANCE JOB SHEET 48

Air Injection System Diagnosis and Service

Name ______________________ Station ______________ Date ______________

NATEF Correlation

This Job Sheet addresses the following NATEF tasks:

E.3.2. Inspect and test mechanical components of secondary air injection systems; perform necessary action.

E.3.3. Inspect and test electrical/electronically-operated components and circuits of air injection systems; perform necessary action.

Objective

Upon completion of this job sheet, you will be able to inspect and test the mechanical components of secondary air injection systems as well as be able to inspect and test the electrical/electronically operated components and circuits of air injection systems.

Tools and Materials

Exhaust gas analyzer

Vacuum gauge

Lab scope

Protective Clothing

Goggles or safety glasses with side shields

Describe the vehicle being worked on:

Year ________ Make ____________________ Model ____________________

VIN ____________________ Engine type and size ____________________

PROCEDURE

Check Valve Testing

1. All of the types of air injection systems have a one-way check valve. The valve opens to let air in but closes to keep exhaust from leaking out. The check valve can be checked with an exhaust gas analyzer. Start the engine and hold the probe of the exhaust gas analyzer near the check valve port. If CO or CO_2 is read, the valve leaks. Summarize the results of this check.

____

____

Pulsed Secondary Air Injection System Diagnosis

1. Check all the hoses and pipes in the system for looseness and rusted or burned conditions. If the metal container or the clean air hose from this container show evidence of burning, some of the one-way check valves are allowing exhaust into this container and clean air hose. Summarize the results of this check.

   ____

   ____

2. Remove the clean air hose from the air cleaner and start the engine. With the engine idling, there should be steady audible pulses at the end of the hose. If these pulses are erratic, check for cylinder misfiring. If the cylinders aren't misfiring, check for sticking one-way check valves or restricted inlet air tubes in the exhaust manifold. Summarize the results of this check.

   ____

   ____

Secondary Air Injection System Diagnosis

1. Check all vacuum hoses and electrical connections in the system. Summarize the results of this check.

   ____

   ____

2. Many AIR system pumps have a centrifugal filter behind the pulley. The pulley and filter are bolted to the pump shaft. These two components are serviced separately. Check the pulley and filter for distortion, wear, contamination, and damage. Summarize the results of this check.

   ____

   ____

3. Check the air pump's belt for condition and tension. What is the tension specification and what tension did you measure?

   ____

4. In some AIR systems, pressure relief valves are mounted in the AIRB and AIRD valves. Other AIR systems have a pressure relief valve in the pump. If the pressure relief valve is stuck open, air flow from the pump is continually exhausted through the valve causing high tail pipe emissions. Does the vehicle have a pressure relief valve? If so, is it stuck open?

   ____

5. Check the hoses in the AIR system for evidence of burning. This would indicate a leak and could be the cause of excessive noise. Summarize the results of this check.

   ____

   ____

6. Check the computer for DTCs related to the AIR system. Some AIR systems will set DTCs if there is a fault in the AIRB or AIRD solenoids and related wiring. In some AIR systems, DTCs are set in the PCM memory if the air flow from the pump is continually upstream or downstream. Use a scan tool to retrieve the codes and summarize the results of this check.

____

____

7. If the HC emissions are high during engine warm-up, check to see if the AIR system pumps air into the exhaust ports during engine warm-up. To check this, see if the PCM remains in open loop for a long time during engine warm-up. Summarize the results of this check.

____

____

8. If the O_2 sensor is always sending a lean signal back to the computer, check the air injection system. Check to see if the air is being pumped into the exhaust ports with the engine at normal operating temperature. The additional air in the exhaust stream causes lean signals from the O_2 sensor and the PCM responds to these lean signals by providing a rich air-fuel ratio from the injectors. Summarize the results of this check.

____

____

Electric AIR Pumps

1. Some late-model AIR systems use an electric air pump controlled by the PCM. These systems have an AIR solenoid and solenoid relay. When the PCM provides a ground for the relay, battery voltage is applied to the solenoid and the pump. Typically, DTCs will be set if one of the components fails or if the hoses or check valves leak. A quick check of the system can be made with a scan tool. Set the scan tool to watch the voltage at the oxygen sensor(s). Task Completed ☐

2. Start the engine and allow it to idle. Task Completed ☐

3. Once the engine has reached normal operating temperature, enable the AIR system and check the HO_2S voltages. If the voltages are low, the AIR pump, solenoid, and shut off valve are working properly. If the voltages are not low, each component of the system needs to be checked and tested. Summarize the results of this check.

____

____

AIR System Component Diagnosis

1. To check an AIRB solenoid and valve, start the engine and listen for air being exhausted from the AIRB valve for a short period of time. If this air is not exhausted, remove the vacuum hose from the AIRB and start the engine. If air is now exhausted from the AIRB valve, check the AIRB solenoid and connecting wires. When air is still not exhausted from the AIRB valve, check the air supply from the pump to the valve. If the air supply is available, replace the AIRB valve. Summarize the results of this check.

   ____

   ____

2. During engine warm-up, remove the hose from the AIRD valve to the exhaust ports and check for air flow from this hose. If air flow is present, the system is operating normally in this mode. When air is not flowing from this hose, remove the vacuum hose from the AIRD valve and connect a vacuum gauge to this hose. If vacuum is above 12 in. Hg, replace the AIRD valve. When the vacuum is zero, check the vacuum hoses, the AIRD solenoid, and connecting wires. Summarize the results of this check.

   ____

   ____

3. With the engine at normal operating temperature, disconnect the air hose between the AIRD valve and the catalytic converters. Check for air flow from this hose. If air flow is present, system operation in the downstream mode is normal. If there is no air flow from this hose, disconnect the vacuum hose from the AIRD valve and connect a vacuum gauge to the hose. If the vacuum gauge indicates no vacuum, replace the AIRD valve. If some vacuum is indicated, check the hose, the AIRD solenoid, and connecting wires. Summarize the results of this check.

   ____

   ____

System Efficiency Test

1. Run the engine at idle with the secondary air system on (enabled). Using an exhaust gas analyzer, measure and record the oxygen (O_2) levels. What were the readings?

   ____

2. Next, disable the secondary air system and continue to allow the engine to idle. Again, measure and record the oxygen level in the exhaust gases. What were the readings?

   ____

3. The secondary air system should be supplying 2 to 5% more oxygen when it is working efficiently. Summarize the results of this check.

   ____

   ____

Problems Encountered

Instructor's Comments

ENGINE PERFORMANCE JOB SHEET 49

Evaporative Emission Control System Diagnosis

Name ______________________ Station ______________ Date ______________

NATEF Correlation

This Job Sheet addresses the following NATEF tasks:

E.4.1. Diagnose emissions and driveability problems resulting from malfunctions in the evaporative emissions control system; determine necessary action.

E.4.2. Inspect and test components and hoses of evaporative emissions control system; perform necessary action.

E.4.3. Interpret evaporative emission related diagnostic trouble codes (DTCs); determine necessary action.

Objective

Upon completion of this job sheet, you will be able to diagnose emissions and driveability problems resulting from the failure of the evaporative emissions control system and you will also be able to inspect and test the components and the hoses of the evaporative emissions control system.

Tools and Materials

Service manual

Digital multimeter

Scan tool

Hand-operated vacuum pump

Protective Clothing

Goggles or safety glasses with side shields

Describe the vehicle being worked on:

Year __________ Make ____________________ Model ____________________

VIN ____________________ Engine type and size ____________________

PROCEDURE

NOTE: *EVAP system diagnosis varies depending on the vehicle make and model year. Always follow the service and diagnostic procedure in the vehicle manufacturer's service manual.*

1. If gasoline odor is present in, or around, a vehicle, check the EVAP system for cracked or disconnected hoses, and check the fuel system for leaks. Gasoline leaks or escaping vapors may result in an explosion, causing personal injury and/or property damage. The cause of fuel leaks or fuel vapor leaks should be repaired immediately. Was there a gas odor? If so, did you find the leak?

____

____

2. If the EVAP system is purging vapors from the charcoal canister when the engine is idling or operating at very low speed, rough engine operation will occur, especially at higher atmospheric temperatures. Does the engine have a rough idle? If so, is the problem canister control?

____

____

3. Check all of the hoses in the EVAP system for leaks, restrictions, and loose connections. Summarize the results of your check.

____

____

4. The electrical connections in the EVAP system should be checked for looseness, corroded terminals, and worn insulation. Summarize the results of your check.

____

____

5. Use a scan tool and check for any EVAP-related DTCs. If a DTC related to the EVAP system is set in the PCM memory, always correct the cause of this code before further EVAP system diagnosis. Summarize the results of your check.

____

____

6. The scan tool may also be used to diagnose the EVAP system. In the appropriate tester mode, the tester indicates whether the purge solenoid is on or off. With the engine idling, the purge solenoid should be off. Leave the scan tool connected and road test the vehicle. Observe the solenoid status on the scan tool. The purge solenoid should be on when all the conditions are present for canister purge operation. If the purge solenoid is not on under the necessary conditions, check the power supply wire to the solenoid, solenoid winding, and the wire from the solenoid to the PCM. Summarize the results of your check.

____

____

7. Check the canister to make sure that it is not cracked or otherwise damaged. Rough idle, flooding, and other conditions can indicate a canister problem. A canister filled with liquid or water causes back-pressure in the fuel tank. It can also cause richness and flooding symptoms during purge or start-up. (Some trucks have intentionally pressurized fuel tank systems. Check the calibration and engine decal before diagnosing.) Summarize the results of your check.

____

____

8. To test for saturation, unplug the canister momentarily and observe the engine's operation. If the canister is saturated, either it or the filter must be replaced depending on its design. That is, some models have a replaceable filter, others do not. Summarize the results of your check.

____

____

9. Check for vacuum leaks in the EVAP system. A vacuum leak in any of the evaporative emission components or hoses can cause starting and performance problems, as can any engine vacuum leak. It can also cause complaints of fuel odor. Summarize the results of your check.

____

____

10. The canister purge solenoid winding may be checked with an ohmmeter. Find the specifications in the service manual and compare your measurements to them. Summarize the results of your check.

____

____

11. With the tank pressure control valve removed, try to blow air through the valve with your mouth from the tank side of the valve. Some restriction to air flow should be felt until the air pressure opens the valve. Summarize the results of your check.

____

____

12. Connect a vacuum hand pump to the vacuum fitting on the valve and apply 10 in. Hg to the valve. Now try to blow air through the valve from the tank side. Under this condition, there should be no restriction to air flow. If the tank pressure control valve does not operate properly, replace the valve. Summarize the results of your check.

____

____

13. If the fuel tank has a pressure and vacuum valve in the filler cap, check these valves for dirt contamination and damage. The cap may be washed in clean solvent. When the valves are sticking or damaged, replace the cap. Summarize the results of your check.

____

____

Problems Encountered

Instructor's Comments

ENGINE PERFORMANCE JOB SHEET 50

Adjusting Valves

Name ______________________________ Station ________________ Date ________________

NATEF Correlation:

This Job Sheet addresses the following NATEF task:

F.1. Adjust valves on engines with mechanical or hydraulic lifters.

Objective

Upon completion of this job sheet, you will be able to adjust the valves on an OHC engine.

Tools and Materials

Service manual

Feeler gauge set

Remote starter button

Hand tools

Protective Clothing

Goggles or safety glasses with side shields

Describe the vehicle being worked on:

Year __________ Make ________________________ Model ________________________

VIN ________________________ Engine type and size ________________________

Describe the valve train arrangement; are there rocker arms, hydraulic lifters, etc.?

____

____

PROCEDURE

NOTE: *This procedure does not apply to all engines and is based on a four-cylinder overhead camshaft engine with valve lash adjusting screws. Engines with different arrangements require a different procedure. Always refer to the service manual for the correct procedure. Record where the manufacturer's procedure varies from the following procedure, at each step.*

1. Check the service manual to determine if the valves should be adjusted with a cold engine or when the engine is at normal operating temperature. Describe the recommended condition.

   ____

   ____

2. What is the specification for valve lash clearance?

____

____

3. When the engine is at the desired temperature, remove the valve (cam) cover.

____

____

4. Set the # 1 piston at top dead center (TDC). Task Completed ☐
5. Select the correct thickness feeler gauge from the set. Task Completed ☐
6. Insert the feeler gauge between the adjusting screw and the end of the valve stem, and slide it back and forth; you should feel a slight amount of drag. What did you observe?

____

____

7. If you felt too much or too little drag, loosen the locknut with the special tools, and turn the adjusting screw until the drag on the feeler gauge is correct. Task Completed ☐
8. Tighten the locknut. Task Completed ☐
9. Recheck the valve clearance. Repeat the adjustment if necessary. Task Completed ☐
10. Check and correct the other valve(s) in cylinder #1. Did you need to adjust them?

____

____

11. Rotate the crankshaft 180° clockwise (the camshaft pulley turns 90°). Task Completed ☐
12. Check and, if necessary, adjust the clearances of the valves on the # 3 cylinder. Did you need to adjust them?

____

____

13. Rotate the crankshaft 180° clockwise (the camshaft pulley turns 90°). Task Completed ☐
14. Check and, if necessary, adjust the clearances of the valves on the # 4 cylinder. Did you need to adjust them?

____

____

15. Rotate the crankshaft 180° clockwise (camshaft pulley turns 90°). Task Completed ☐

16. Check and, if necessary, adjust the clearances of the valves on the # 2 cylinder. Did you need to adjust them?

____

____

Task Completed ☐

17. Install the cylinder head cover.

Problems Encountered

____

____

____

Instructor's Comments

____

____

____

ENGINE PERFORMANCE JOB SHEET 51

Replace a Timing Belt on an OHC Engine

Name ______________________ Station ______________ Date ______________

NATEF Correlation

This Job Sheet addresses the following NATEF task:

F.2. Remove and replace timing belt; verify correct camshaft timing.

Objective

Upon completion of this job sheet, you will be able to inspect and replace timing belt(s), overhead cam drive sprockets, and tensioners, as well as check belt tension.

Tools and Materials

Basic hand tools

Paint stick or chalk

Belt tension gauge

Protective Clothing

Goggles or safety glasses with side shields

Describe the vehicle being worked on:

Year __________ Make ____________________ Model ____________________

VIN ____________________ Engine type and size ____________________

PROCEDURE

1. Disconnect the negative cable from the battery prior to beginning to remove and replace the timing belt.
2. Carefully remove the timing cover, being careful not to distort or damage it while pulling it up. With the cover removed, check the immediate area around the belt for wires and other obstacles. If some are found, move them out of the way. What needed to be removed? Did you need to remove other drive belts?

   ____

   ____

   ____

3. Align the timing marks on the camshaft's sprocket with the mark on the cylinder head. If the marks are not obvious, use a paint stick or chalk to clearly mark them. Task Completed ☐
4. Carefully remove the crankshaft timing sensor and probe holder. Task Completed ☐

5. Loosen the adjustment bolt on the belt tensioner pulley. It is normally not necessary to remove the tensioner assembly. Task Completed ☐

6. Slide the belt off the crankshaft sprocket. Do not allow the crankshaft pulley to rotate while doing this. Task Completed ☐

7. To remove the belt from the engine, the crankshaft pulley may need to be removed to slip it off the crankshaft sprocket. Did you need to remove the pulley? What did you need to do in order to do this?

 ____

 ____

 ____

8. After the belt has been removed, inspect it for cracks and other damage. Cracks will become more obvious if the belt is twisted slightly. Describe any defects in the belt. **NOTE:** *Timing belts are always replaced once they have been removed. When may this not be true?*

 ____

 ____

 ____

9. To begin reassembly, place the belt around the crankshaft sprocket. Then reinstall the crankshaft pulley. Task Completed ☐

10. Make sure the timing marks on the crankshaft pulley are lined up with the marks on the engine block. If they are not, carefully rock the crankshaft until the marks are lined up. Task Completed ☐

11. With the timing belt fitted onto the crankshaft sprocket and the crankshaft pulley tightened in place, the crankshaft timing sensor and probe can be reinstalled. Task Completed ☐

12. Align the camshaft sprocket with the timing marks on the cylinder head. Then wrap the timing belt around the camshaft sprocket and allow the belt tensioner to put a slight amount of pressure on the belt. Task Completed ☐

13. Adjust the tension as described in the service manual. Then rotate the engine two complete turns. Recheck the tension. What are the specifications for belt tension? Why do you need to rotate the engine twice before rechecking the tension?

 ____

 ____

 ____

 ____

14. Rotate the engine through two complete turns again, then check the alignment marks on the camshaft and the crankshaft. Any deviation needs to be corrected before the timing cover is reinstalled. Task Completed ☐

Problems Encountered

Instructor's Comments

ENGINE PERFORMANCE JOB SHEET 52

Servicing a Thermostat

Name ______________________ Station ______________ Date ______________

NATEF Correlation

This Job Sheet addresses the following NATEF task:

F.3. Remove and replace thermostat and gasket.

Objective

Upon completion of this job sheet, you will be able to inspect and test the thermostat, by-pass, and the housing.

Tools and Materials

Thermometer
Container for water
Heat source for heating the water
Basic hand tools

Protective Clothing

Goggles or safety glasses with side shields

Describe the vehicle being worked on:

Year __________ Make ______________________ Model ______________________

VIN ______________________ Engine type and size ______________________

PROCEDURE

1. Thoroughly inspect the area around the thermostat and its housing. Describe the result of that inspection.

____

____

____

2. There are several ways to test the opening temperature of a thermostat. The first method does not require that the thermostat be removed from the engine. Begin by removing the radiator pressure cap from a cool radiator and insert a thermometer into the coolant. Task Completed ☐

3. Start the engine and let it warm up. Watch the thermometer and the surface of the coolant. When the coolant begins to flow or move in the radiator, what is indicated?

____

____

4. When the fluid begins to flow, record the reading on the thermometer. If the engine is cold and coolant circulates, this indicates the thermostat is stuck open and must be replaced. The measured opening temperature of the thermostat was: ____________________

5. Compare the measured opening temperature with the specifications. The specifications call for the thermostat to open at what temperature? __________ What do you recommend based on the results of this test?

____

____

6. The other way to test a thermostat is to remove it. Begin removal by opening the radiator pressure cap to relieve pressure, if the cap is still on. Task Completed ☐

7. Drain coolant from the radiator until the level of coolant is below the thermostat housing. Recycle the coolant according to local regulations. What did you do with the drained coolant?

____

____

____

8. Disconnect the upper radiator hose from the thermostat housing. Task Completed ☐

9. Unbolt and remove the housing from the engine. The thermostat may come off with the housing. Task Completed ☐

10. Thoroughly clean the gasket surfaces for the housing. Make sure the surfaces are not damaged while doing so and that gasket pieces do not fall into the engine at the thermostat bore. Task Completed ☐

11. Carefully inspect the thermostat housing. Describe your findings.

____

____

12. Suspend the thermostat while it is completely submerged in a small container of water. Make sure it does not touch the bottom of the container. What did you use to suspend it?

____

____

____

13. Place a thermometer in the water so that it does not touch the container and only measures water temperature. Task Completed ☐

14. Heat the water. When the thermostat valve barely begins to open, read the thermometer. What was the measured opening temperature of the thermostat? Compare that with the specifications.

____

____

15. Remove the thermostat from the water and observe the valve. Record what happened and your conclusions about the thermostat.

____

____

____

16. Carefully look over the new (or old if okay) thermostat. Identify which end of the thermostat should face toward the radiator. How did you determine the proper direction for installation?

____

____

17. Fit the thermostat in the recessed area in the engine or thermostat housing. Where was the recess? ____________________________

18. Refer to the installation instructions on the gasket's container. Should an adhesive and/or sealant be used with the gasket? If so, what?

____

____

19. Install the gasket according to the instructions. Task Completed ☐

20. Install the thermostat housing. Before tightening it in place, make sure it is fully seated and flush onto the engine. Failure to do this will result in a broken housing. Task Completed ☐

21. Tighten the bolts evenly and carefully and to the correct specifications. What are the specifications? ____________________________

22. Connect the upper radiator hose to the housing with a new clamp. Why should you use a new clamp?

____

____

____

23. Pressurize the system and check for leaks. Why should you do this now?

____

____

24. Replenish the coolant and bring it to its proper level. Install the radiator cap. Task Completed ☐

25. Run the engine until it is at normal operating temperature. Did the cooling system warm up properly and does it seem that the thermostat is working properly?

____

____

26. Recheck the coolant level. Task Completed ☐

Problems Encountered

____

____

____

Instructor's Comments

____

____

____

ENGINE PERFORMANCE JOB SHEET 53

Checking Cooling Fans

Name ______________________ Station ______________ Date ______________

NATEF Correlation

This Job Sheet addresses the following NATEF task:

F.4. Inspect and test mechanical/electrical fans, fan clutch, fan shroud/ducting, air dams, and fan control devices; perform necessary action.

Objective

Upon completion of this job sheet, you will be able to inspect and test mechanical/electrical fans, the fan clutch, fan shroud/ducting, air dams, and the fan control devices.

Tools and Materials

A vehicle with a mechanical cooling fan

A vehicle with an electrical cooling fan

Thermometer

Ignition timing light

Tachometer

Protective Clothing

Goggles or safety glasses with side shields

Mechanical Fan

Describe the vehicle being worked on:

Year __________ Make ____________________ Model ____________________

VIN ____________________ Engine type and size ____________________

PROCEDURE

1. Inspect the fan shroud and air dams. Make sure they are firmly mounted to the radiator and/or frame of the vehicle. Summarize your findings.

   ____

   ____

2. Check the straightness of the fan blades. Summarize your findings.

   ____

   ____

3. Check for missing or damaged attaching rivets. Check the attaching point for each fan blade. Summarize your findings.

____

____

4. Check the condition of the drive pulley for the fan. Summarize your findings.

____

____

5. If the fan is equipped with a clutch, continue this job sheet. Task Completed ☐

6. Use your hands to spin the fan. If the fan rotates more than twice, the clutch is bad. Summarize your findings.

____

____

7. Check for oil around the clutch. Oil indicates that the viscous clutch is leaking and should be replaced. Summarize your findings.

____

____

8. Attach a thermometer to the engine side of the radiator. Task Completed ☐

9. Connect a timing light and tachometer to the engine. Task Completed ☐

10. Start the engine, then observe and record the reading on the thermometer when the engine is cold. Task Completed ☐

11. Aim the timing light at the fan blades; they should appear to being moving very slowly. Summarize your findings.

____

____

12. Put a large piece of cardboard across the radiator to block air flow. Watch the thermometer and the timing light flashes as the engine warms up. Summarize your findings.

____

____

13. Refer to the service manual to find the specified fan engagement temperature. What is it?

____

____

14. Once the engine reaches that temperature, remove the cardboard. The fan clutch should now cause the fan speed to increase. Summarize your findings.

____

____

Electric Fan

Describe the vehicle being worked on:

Year __________ Make ________________________ Model ________________________

VIN ________________________ Engine type and size ________________________

PROCEDURE

1. Visually inspect the electric cooling fan(s) and related wiring to determine if they are clean and properly connected. Task Completed ☐
2. Run the engine until it reaches normal operating temperature to determine if the electric cooling fan is working. Task Completed ☐
3. If the fan is operating properly, blow the fan and surrounding area clean with air. Task Completed ☐
4. If the fan is not working, check the power wire to the fan to determine if there is power. Task Completed ☐
5. Check the ground circuit to determine if the system has a good ground. Task Completed ☐
6. If power and a good ground are present, remove the electric fan assembly and replace it. Task Completed ☐
7. If there is no power, check the fuse and fan relay. The location of the fuse and relay can be found in the service manual. Task Completed ☐
8. Replace bad components, then retest fan operation. Task Completed ☐
9. If there is not a good ground, check the temperature sensor that controls the electric fan. The location of the sensor and the proper test procedure can be found in the service manual. Task Completed ☐
10. If the sensor fails the test, replace it. Check the operation of the electric fan. Task Completed ☐

WARNING: *The electric cooling fans can operate with the engine and the key both turned off. The fan can come on at any time if the engine is hot or there is a problem with a sensor. Keep your hands away from the fan at all times.*

Problems Encountered

____

____

____

Instructor's Comments

____

____

____

ENGINE PERFORMANCE JOB SHEET 54

How to Repair and Replace Damaged Threads

Name ______________________ Station ______________ Date ______________

NATEF Correlation:

This Job Sheet addresses the following NATEF task:

F.5. Perform common fastener and thread repair to include, remove broken bolt, restore internal and external threads, and repair internal threads with thread insert.

Objective

Upon completion of this job sheet, you will be able to inspect and repair internal and external threads, as well as install thread inserts and remove broken bolts.

Tools and Materials

Vise grips
Small cape chisel
Screw extractor set
Tap and die set
Drill
Drill bits
Center punch
Cutting fluid
Air nozzle
Heli-coil or TimeSert set

Protective Clothing

Goggles or safety glasses with side shields

PROCEDURE

Removing Broken Bolts

1. If there is part of the broken bolt sticking above the surface, grip that part with vise grips and attempt to turn the bolt out. Were you able to do this?

____

____

NOTE: *Applying heat with a pencil/pinhead torch to area around the broken bolt will help loosen it.*

2. If vise grips did not work or if the bolt is broken even with or below the surface, place a small cape chisel on the broken surface of the bolt. Task Completed ☐

3. Use a hammer to tap the chisel and bolt to turn the bolt out. Once enough of the bolt is exposed, use the vise grips to remove it. Were you able to do this?

____

____

4. If you still cannot get the bolt out, get a screw extractor set. Task Completed ☐

5. With a center punch, and make a dimple in the exact center of the broken bolt. Task Completed ☐

6. With the correct size drill bit (as recommended by the manufacturer of the screw extractor set), drill through the center of the bolt to the depth recommended by the tool manufacturer. Take care not to drill too deeply and make sure the drill bit moves straight through the center of the bolt. Task Completed ☐

7. Install the correct extractor. Most often, the extract will need to be tapped with a hammer into the new bore; other extractors have left-hand threads and must be threaded into the bore. Task Completed ☐

8. Turn the extractor to remove the bolt. Did the bolt come out?

____

____

9. After the bolt has been removed, run a thread chaser through the bolthole to straighten and clean the threads in the bore. Task Completed ☐

Restoring Internal Threads

1. Using a thread gauge, determine the proper thread pitch and size of the damaged threads. Describe the threads.

____

____

2. Select the correct tap for the bolthole. Task Completed ☐

3. Apply some cutting fluid to the tap. Task Completed ☐

4. Turn the tap slowly into the hole. Stop at about a quarter to a half turn. Then back the tap out to keep the cutting edges of the tap clean. Task Completed ☐

5. Continue the process until the tap hits the bottom or the required depth needed. Then turn the tap totally out. Task Completed ☐

6. Use compressed air to clean out the hole and use a solvent to remove the cutting oil. Task Completed ☐

7. Run the correct size thread chaser through the threads and clean out the bore with compressed air. Task Completed ☐

8. Explain any problems you had in doing this.

____

____

Restoring External Threads

1. Using a thread gauge, determine the proper thread pitch and size of the damaged threads. Describe the threads.

____

____

2. Thoroughly clean the threads that will be repaired. Task Completed ☐
3. Apply some cutting fluid to the die. Task Completed ☐
4. Carefully start the die over the threads. If the outer edge of the threads is damaged, it is helpful to file the sharp edges off before starting the die. Task Completed ☐
5. Turn the die slowly over the threads. Stop at about a quarter to a half turn. Then reverse the die to keep the cutting edges of the die clean. Task Completed ☐
6. Continue the process until the die reaches the end of the threads. Then turn the die off the bolt. Task Completed ☐
7. Clean the bolt and die with compressed air and solvent. Task Completed ☐
8. Explain any problems you had in doing this.

____

____

Installing Thread Inserts

1. Determine the correct size and thread pitch for the bore. What is it?

____

____

2. Select the appropriate insert from the kit. Task Completed ☐
3. Using the guidelines of the insert manufacturer, determine the drill size required to install the insert. What size will you use?

____

____

4. While making sure the drill is perfectly straight, drill the required bore to the required depth. If you are going to install a TimeSert, put the required chamfer on the outside of the bore. Task Completed ☐
5. Clean out that bore with compressed air. Task Completed ☐
6. With the correct tap, thread the inside of the bore to match the external threads of the insert. Task Completed ☐
7. Screw the insert tightly into the bore. On HeliCoil type inserts, the tang at the top of the insert must be removed with needle nosed pliers after the insert is fully seated. Task Completed ☐
8. Explain any problems you had in doing this.

____

____

Problems Encountered

Instructor's Comments

ENGINE PERFORMANCE JOB SHEET 55

Perform Oil and Filter Change

Name ______________________ Station ______________ Date ______________

NATEF Correlation

This Job Sheet addresses the following NATEF tasks:

F.6. Perform oil and filter changes.

Objective

Upon completion of this job sheet, you will be able to properly change an engine's oil and oil filter.

Tools and Materials

Rags
Funnel
An oil filter wrench
Lift

Protective Clothing

Goggles or safety glasses with side shields

Describe the vehicle being worked on:

Year ________ Make ____________________ Model ____________________

VIN ____________________ Engine type and size ____________________

PROCEDURE

1. Always make sure the vehicle is positioned safely on a lift or supported by jack stands. Before raising the vehicle, allow the engine to run awhile. After it is warm, turn off the engine. Task Completed ☐
2. Place the oil drain pan under the drain plug before beginning to drain the oil. Task Completed ☐
3. Loosen the drain plug with the appropriate wrench. After the drain plug is loosened, quickly remove it so the oil can freely drain from the oil pan. Make sure the drain pan is positioned so it can catch all of the oil. Describe the color and condition of the oil.

   ____

   ____

4. While the oil is draining, use an oil filter wrench to loosen and remove the oil filter. Describe what you needed to do in order to do this.

____

____

5. Make sure the oil filter seal came off with the filter. Then place the filter into the drain pan so it can drain. After it has completely drained, discard the filter according to local regulations. What are your local requirements for disposing oil filters and oil?

____

____

6. Wipe off the oil filter sealing area on the engine block. Then apply a coat of clean engine oil onto the new filter's seal. Task Completed ☐

7. Install the new filter and hand-tighten it. Oil filters should be tightened according to the directions given on the filter. What are the instructions?

____

____

8. Prior to installing the drain plug, wipe off its threads and sealing surface with a clean rag. Inspect the threads and describe their condition:

____

____

9. The drain plug should be tightened according to the manufacturer's recommendations. Overtightening can cause thread damage, while undertightening can cause an oil leak. What is the tightening spec for the plug?

____

10. With the oil filter and drain plug installed, lower the vehicle and remove the oil filler cap. Task Completed ☐

11. Carefully pour the oil into the engine. The use of a funnel usually keeps oil from spilling on the engine. What type of oil is recommended and how much?

____

12. After the recommended amount of oil has been put in the engine, check the oil level. Task Completed ☐

13. Start the engine and allow it to reach normal operating temperature. While the engine is running, check the engine for oil leaks, especially around the oil filter and drain plug. If there is a leak, shut down the engine and correct the problem. Were there any leaks? Where?

____

____

14. After the engine has been turned off, recheck the oil level and correct it as necessary. Task Completed ☐

Problems Encountered

___

___

___

Instructor's Comments

___

___

___

ENGINE PERFORMANCE JOB SHEET 56

Using an Oxy-Acetylene Torch

Name ______________________ Station ______________ Date ______________

NATEF Correlation

This Job Sheet addresses the following NATEF task:

F.7. Demonstrate proficiency in using oxy-acetylene torch to heat and cut metal.

Objective

Upon completion of this job sheet, you will be able to heat and cut metal with an oxy-acetylene torch.

Tools and Materials

Oxygen cylinder
Acetylene cylinder
Pressure regulators for both gases
Oxygen gauge that reads up to 30 lb/square inch
Acetylene gauge that reads up to 15 lb/square inch
Welding torch with various tips
Hoses
Igniter (sparker)
Cutting attachment
Welder's chalk
Chipping hammer
Wire brush

Protective Clothing

Welding goggles
Leather apron
Leather arm coverings
Leather gloves
Steel toe leather boots

Describe the vehicle being worked on:

Year __________ Make ______________________ Model ______________________

VIN ______________________ Engine type and size ______________________

PROCEDURE

Precautions

1. Read through the following precautions before proceeding to set up the equipment and use it. Task Completed ☐
   - Do not watch the torch's flame with your eyes unprotected by a sufficiently darkened glass.
   - Always wear protective gloves, aprons, arm protection.
   - Never wear frayed clothing that could be caught on fire by sparks.

- Welding galvanized steel is very dangerous to one's health; the galvanized coating on sheet metal has a zinc component in it and this is very bad to breathe in – it can cause serious health problems.
- Acetylene is an extremely flammable gas.
- Never use an oxy-acetylene torch in a confined space without proper ventilation.
- NEVER use oil on the threads of a gas cylinder when hooking up gauges. This is a serious explosion hazard.
- Never use a regulator on any gas except that for which it was designed.
- Do not use regulators with broken gauges.
- When a gas cylinder is in use, it must be properly secured to a cart.
- When a cylinder is being moved, it must have the cap screwed on to protect the valve from accidental damage.
- When a cylinder is in storage before or after use, it must be properly secured to prevent it from falling.
- Make sure the hoses are not kinked or damaged.
- Make sure the torch is shut off when it is not in use.

Getting the Equipment Ready

1. Move the gas cylinders to the area where you will be working. Identify the type of gas by the color of the cylinder and state the color of each.

   ____

   ____

2. Carefully inspect the valve on the cylinders. Make sure they are free from oil and grease. Task Completed ☐
3. Stand to the side of the cylinders and make sure the valves are pointing away from you. Also, make sure that any gas coming out of the valves will not hit any oily material. Task Completed ☐
4. Open the valves slightly for an instant and then close them. This removes any dust or foreign particles from the area where the gauges are to be attached. Task Completed ☐
5. Screw the regulator into the valve on the cylinder until it is seated. Do not use excessive force, but make sure the joints are gas-tight. The two cylinders have different thread designs; describe the threads of the oxygen and acetylene cylinders.

   ____

   ____

6. Connect the hose assembly to the screwed outlets of the regulators. Again, the hoses are threaded differently. Task Completed ☐
7. Connect the other end of the hose to the torch handle. Jeep the valves on the handle closed while doing this. Task Completed ☐

8. Connect the correct size torch tip to the end of the torch handle. Refer to the torch's manufacturer for the recommended tip for the job and for the recommended pressures to use. Since the first operation in the job sheet is heating metal, connect the appropriate cutting torch and tip. A welding tip can also be used to heat and can now be installed if desired. Task Completed ☐

9. Open the valves on the torch. Task Completed ☐

10. Open the valves on the cylinders. Task Completed ☐

11. Set the regulators to the desired pressures. What pressures are you going to use?

____

____

12. After the pressures are set, close the torch valves. Task Completed ☐

13. With a soap and water solution, check the system for leaks. Apply the solution at all connections using a small soft brush. What did you find?

NOTE: *When a leak is found, immediately turn off the cylinders and correct the problem. Never use jointing compounds or sealing tapes to correct leaks.*

____

____

Lighting

1. Open the cylinder valves, no more than one turn, slowly with the cylinder key. Do not open these quickly. Doing so may cause damage to the regulator. Task Completed ☐

2. Open the fuel gas control valve on the torch and adjust the regulator to the correct working pressure. Task Completed ☐

3. Repeat the same procedure for the oxygen side. Task Completed ☐

4. Locate the striker and make sure it makes a spark. Task Completed ☐

5. Make sure the valves are closed on the torch head. Task Completed ☐

6. Point the torch in a safe direction before lighting. Task Completed ☐

7. Stand to the side of the gauges and open the acetylene cylinder slowly, about 1/4 turn. Never open an acetylene cylinder more than 3/4 turn, and never operate acetylene at pressures greater than 15 psi. Task Completed ☐

8. Using the valves at the torch, purge both oxygen and fuel from the lines prior to lighting the torch. Task Completed ☐

9. Stand to the side of the gauges and slowly open the oxygen cylinder and eventually open it all the way. Task Completed ☐

10. Make sure you are wearing welding gloves and goggles, then slightly open the acetylene valve on the torch and use the striker to ignite the acetylene. Task Completed ☐

11. Adjust this flame until you have a good flame at the desired pressure. Task Completed ☐

12. Slowly open the oxygen valve on the torch and adjust it until you have a neutral flame. Describe what a neutral flame looks like.

____

____

13. Use the torch and flame as needed, and then shut off the acetylene valve. Task Completed ☐
14. Then shut off the oxygen valve. Task Completed ☐
15. Close the valves on the cylinders. Task Completed ☐
16. Open the valves on the torch to bleed the lines. Do this, one at a time, until the gauges drop to zero. Task Completed ☐
17. Back off the regulator adjusting screws to make sure there is no pressure on the diaphragm. Task Completed ☐

Heating

1. Using a torch to heat something is typically done to loosen a stubborn bolt or nut. What component or bolt/nut will you be using for this job sheet?

____

____

2. When attempting to remove a nut, set the flame on the torch. Task Completed ☐
3. Apply the heat of the flame to the nut. Try to limit the amount of heat that moves to the stud or bolt that the nut is on. Task Completed ☐
4. Once the nut is very red, use a wrench or vise grips to remove it. Were you successful?

____

____

5. When attempting to remove a stuck bolt, apply the heat of the flame to the area surrounding the bolt. However, apply only as much heat as necessary. The heat can distort the component. Task Completed ☐

Cutting

1. Make sure the object to be cut is secure and easy to access. What will you be cutting?

____

____

2. Make a line where the metal needs to be cut with welder's chalk. Task Completed ☐
3. Install the appropriate cutting tip into the torch handle. These tips come in a variety of sizes and styles for the various gasses and material thicknesses. Task Completed ☐
4. Open the cylinder valves and purge the lines. Task Completed ☐

5. Set the working oxygen pressure on the regulator with the oxygen passing through the cutting oxygen valve on the cutter. Task Completed ☐
6. Turn off the valves on the torch. Task Completed ☐
7. Open acetylene valve about a half turn and light it with the sparker. Task Completed ☐
8. Slowly open the oxygen valve and adjust the flame to a bright inner cone, but no feathers. Task Completed ☐
9. Depress cutting lever and adjust the flame so it has a bright inner cone. Task Completed ☐
10. Slowly open the oxygen valve on the cutter and adjust the flame to neutral. Task Completed ☐
11. Now depress the cutting oxygen lever and again adjust the controls to give a neutral flame. Task Completed ☐
12. Preheat the metal edge to a dull cherry red. A tight circular pattern over the starting area will even out the preheating. Task Completed ☐
13. Heat the cutting area on the metal until it is cherry red. If the metal is too cool, it will not cut. If it is too hot, as in white hot, it splatters, undercuts, and reduces the quality of the cut. Task Completed ☐
14. When a small cherry red dot appears, pull up on the oxygen cutting valve lever. Task Completed ☐
15. Hold the torch at a vertical angle and slowly follow the cut line with the torch until the cut is complete. Task Completed ☐
16. Cool the metal thoroughly. Task Completed ☐
17. Clean the cuts with chipping hammer and wire brush. Task Completed ☐

Closing Down

1. When you are finished using the torch, turn the acetylene valve on the torch off first. Task Completed ☐
2. Then, turn the oxygen valve off. Task Completed ☐
3. Close the cylinder valves. Task Completed ☐
4. Release the pressure in the hoses by first opening the oxygen valve on the torch and then closing it. Task Completed ☐
5. Then open the acetylene valve and close it. Task Completed ☐
6. Unscrew the pressure regulating screws on the oxygen and acetylene regulators. Task Completed ☐

Problems Encountered

Instructor's Comments

ENGINE PERFORMANCE JOB SHEET 57

Servicing the Internal Combustion Engine on a Hybrid Vehicle

Name ______________________ Station ______________ Date ______________

NATEF Correlation

This Job Sheet addresses the following NATEF task:

F.8. Identify hybrid vehicle internal combustion engine service precautions.

Objective

Upon completion of this job sheet, you will be able to locate and safely disconnect the high voltage circuit on a hybrid vehicle. This will allow you to safely work on the internal combustion engine in this vehicle.

Tools and Materials

Service manual

Hybrid vehicle

Vinyl tape for insulation

Protective Clothing

Goggles or safety glasses with side shields

Insulated gloves that are dry and are not cracked, ruptured, torn, or damaged in any way.

Describe the vehicle being worked on:

Year __________ Make ____________________ Model ____________________

VIN ____________________ Engine type and size ____________________

PROCEDURE

WARNING: *Unprotected contact with any electrically charged ("hot" or "live") high-voltage component could cause serious injury or death.*

1. Describe how this vehicle is identified as a hybrid?

____

____

____

2. After referring to the vehicle's owner's manual, briefly describe its operation.

____

____

____

____

3. How many volts are the batteries rated at?

____

4. Where are the batteries located?

____

____

5. How are the high-voltage cables labeled and identified?

____

____

6. What is used to provide short circuit protection in the high-voltage battery pack?

____

____

7. What isolates the high-voltage system from the rest of the vehicle when the vehicle is shut off?

____

____

8. Using the vehicle's service and/or owner's manual as a guide, list at least five precautions that must be adhered to when servicing this hybrid vehicle?

____

____

____

____

____

9. If the vehicle has been in an accident and some electrolyte from the batteries has leaked out, how should you clean up the spill?

____

____

____

____

10. Look under the hood and describe the components that are visible.

____

____

____

____

11. Without touching the high-voltage cables or components, describe the routing of the cables. Include what they appear to be connected from and to.

____

____

____

____

WARNING: *Never assume that a hybrid vehicle is shut off simply because it is silent. Make sure the ignition key is in your pocket and not in the ignition switch.*

12. Describe the procedure for totally isolating the high-voltage system from the vehicle. This often involves the removal of a service plug. Be sure to include the location of this plug in your description of the procedure.

____

____

____

____

____

____

____

____

____

____

13. Make two signs saying, "*WORKING ON HIGH VOLTAGE PARTS. DO NOT TOUCH!*" Attach one to the steering wheel, and set the other one near the parts you are working on. Task Completed ☐

14. After the service plug has been removed, how long should you wait before working around or on the high-voltage system?

____

____

Problems Encountered

____

____

____

Instructor's Comments

____

____

____

— NOTES —

— NOTES —

— NOTES —

— NOTES —

— NOTES —

— NOTES —

— NOTES —

— NOTES —

— NOTES —